Discovering the World of Geography
Grades 6–7

By
MYRL SHIREMAN

COPYRIGHT © 2003 Mark Twain Media, Inc.

ISBN 1-58037-229-5

Printing No. CD-1575

Mark Twain Media, Inc., Publishers
Distributed by Carson-Dellosa Publishing Company, Inc.

Table of Contents

Introduction

The eighteen National Geography Standards identify the skills and knowledge needed by United States students. The standards are designed to make students in the United States competitive on an international basis.

This book is written to support teachers' efforts to help students in grades six or seven become more competitive on an international level. Because the book is only 128 pages, it has been necessary to design learning activities that are teacher-friendly and develop a selected skill and knowledge base that can be applied to selected performance standards. Additional skills, knowledge, and performance standards will be incorporated in a book on the Eastern Hemisphere for grades seven or eight.

The first nine chapters are devoted to building the students' skills and knowledge as they apply to the nations in the Western Hemisphere. Student activities, followed by pretest practice and a test format, are used to ensure student mastery of a solid skill and knowledge base. The pretests and final tests are designed to be administered following students' completion of each chapter. Individual teachers may choose to use the tests in a different manner. The pretest is designed to prompt students to the correct answer. The final test for each chapter is in a traditional multiple-choice format.

To successfully complete Chapters 10, 11, and 12, students must apply the skills and knowledge learned in the first nine chapters. These chapters are designed to develop each student's ability to question the relationship among people, places, and the environment. These chapters will hopefully help students become more competent in geography by learning to ask appropriate questions, analyze geographic information, present geographic information, and to form generalizations.

National Geography Standards

Teachers leading discussions while completing units and activities is a prerequisite for accomplishing the standards. Inquiry discussion is important: *What do you know? How are things alike and not alike? Why do you think that?*

National Standard 1: *How to use maps and other geographic representations, tools, and technologies to acquire, process, and report information*

This standard is addressed through map, chart, and graph activities in a number of chapters. Unit 4 and Unit 12: Problems 1, 4, and 13 specifically address this standard.

National Standard 4: *The physical and human characteristics of place*

This standard is addressed in activities throughout the book. Unit 10, Unit 11, and Unit 12: Problems 2, 7, and 11 specifically address this standard.

National Standard 9: *The characteristics, distributions, and migrations of human populations on Earth's surface*

Activities in Unit 2 address this standard as students learn about the populations of Western Hemisphere nations. Unit 12: Problem 10 specifically addresses this standard.

National Standard 10: *The character, distribution, and complexity of Earth's cultural mosaics*

This standard is specifically addressed in Unit 12: Problem 13.

National Standard 11: *The patterns and networks of economic interdependence on Earth's surface*

This standard is addressed in Unit 7 and Unit 8 as students learn of the agriculture and natural resources of Western Hemisphere nations. Unit 12: Problems 8 and 9 specifically address this standard.

National Standard 12: *The process, patterns, and functions of human settlement*

Throughout the book, students address this standard as they learn how physical features, natural resources, agriculture, and climate affect human settlement. The standard is specifically addressed in Unit 12: Problems 5 and 6.

National Standard 15: *How physical systems affect human systems*

This standard is addressed in Units 3 and 5 as students learn how physical features affect the climate and the desirability of a region for settlement, transportation, commerce, and agriculture. Unit 12: Problems 3, 6, and 12 specifically address this standard.

Name: _____ Date: _____

Unit 1: Political Geography of Western Hemisphere Nations

A. Western Hemisphere

Using **Map 1** and **an atlas**, complete the following. You will need colored pens or pencils for this activity.

1. The number 1 locates Canada. Color it blue.
2. The number 2 locates the United States. Color it red.
3. The number 3 locates Mexico. Color it yellow.
4. The number 4 locates Central America. Color it orange.
5. The number 5 locates the West Indies (Caribbean Islands). Color it brown.
6. The number 6 locates South America. Color it purple.

MAP 1

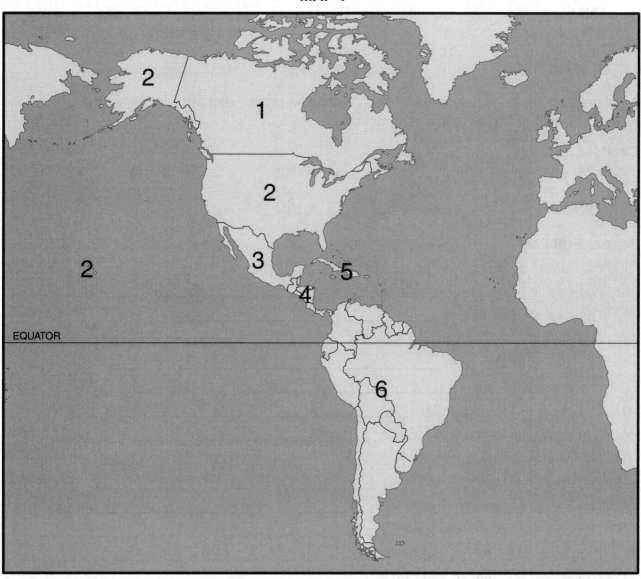

Name: _____   Date: _____

Unit 1: Political Geography of Western Hemisphere Nations

B. Canadian Provinces/Territories and Capitals

Canada is a nation made up of provinces and territories. Each province and territory has a capital city.

Using **Map 2** and **an atlas**, complete the following about Canada's provinces and territories. Place the number by the province or territory on the map to show its location.

Provinces and Territories
1. British Columbia
2. Ontario
3. Prince Edward Island
4. Alberta
5. Quebec
6. Nova Scotia
7. Saskatchewan
8. Newfoundland
9. Manitoba
10. New Brunswick
11. Yukon Territory
12. Northwest Territories
13. Nunavut Territory

A dot on **Map 2** locates the capital of each Canadian province or territory. Place the letter of the capital city next to the corresponding dot on the map.

Capital Cities
a. Charlottetown
b. St. John's
c. Halifax
d. Fredericton
e. Quebec
f. Toronto
g. Winnipeg
h. Regina
i. Edmonton
j. Victoria
k. Whitehorse
l. Yellowknife
m. Iqaluit

Pretest Practice

Use the names of the above capital cities to fill in the blanks.

1. The capital of British Columbia is _____.
2. The capital of Ontario is _____.
3. The capital of Prince Edward Island is _____.
4. The capital of Alberta is _____.
5. The capital of Quebec is _____.
6. The capital of Nova Scotia is _____.
7. The capital of Saskatchewan is _____.
8. The capital of Newfoundland is _____.
9. The capital of Manitoba is _____.
10. The capital of New Brunswick is _____.
11. The capital of the Yukon Territory is _____.
12. The capital of the Northwest Territories is _____.
13. The capital of Nunavut Territory is _____.

Name: _____ Date: _____

Unit 1: Political Geography of Western Hemisphere Nations

B. Canadian Provinces/Territories and Capitals (cont.)

MAP 2

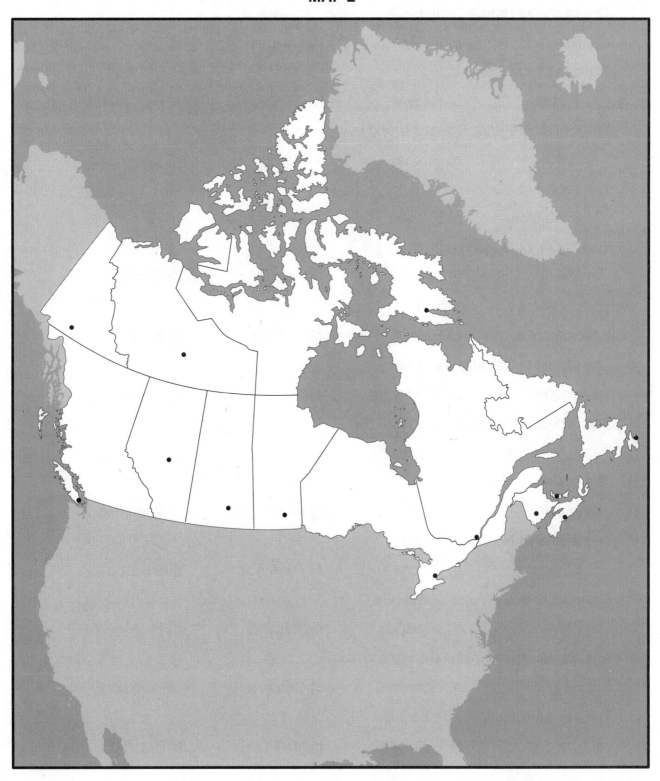

Name: _____ Date: _____

Unit 1: Political Geography of Western Hemisphere Nations

B. Canadian Provinces/Territories and Capitals—Test

Circle the letter of the correct answer.

1. The capital of British Columbia is

    a) Edmonton   b) Victoria   c) Quebec   d) Yellowknife.

2. The capital of Ontario is

    a) Toronto   b) Regina   c) Quebec   d) Fredericton.

3. The capital of Prince Edward Island is

    a) Iqaluit   b) Toronto   c) Halifax   d) Charlottetown.

4. The capital of Alberta is

    a) Edmonton   b) Victoria   c) Whitehorse   d) Yellowknife.

5. The capital of Quebec is

    a) Quebec   b) Regina   c) Iqaluit   d) Toronto.

6. The capital of Nova Scotia is

    a) Edmonton   b) Halifax   c) Quebec   d) St. John's.

7. The capital of Saskatchewan is

    a) Quebec   b) Winnipeg   c) Iqaluit   d) Regina.

8. The capital of Newfoundland is

    a) Yellowknife   b) Regina   c) Victoria   d) St. John's.

9. The capital of Manitoba is

    a) Iqaluit   b) St. John's   c) Halifax   d) Winnipeg.

10. The capital of New Brunswick is

    a) Edmonton   b) Victoria   c) Fredericton   d) Quebec.

11. The capital of the Yukon Territory is

    a) Whitehorse   b) Winnipeg   c) Iqaluit   d) Regina.

12. The capital of the Northwest Territories is

    a) Regina   b) Victoria   c) Quebec   d) Yellowknife.

13. The capital of Nunavut Territory is

    a) Quebec   b) Winnipeg   c) Iqaluit   d) Regina.

Name: _____ Date: _____

Unit 1: Political Geography of Western Hemisphere Nations

C. United States and Capitals

The United States includes fifty states and the commonwealth of Puerto Rico. Each state or commonwealth has a capital city.

Using **Map 3** and **an atlas**, place the number by each state or commonwealth on the map to show its location.

States

1. Alaska	2. Hawaii	3. Texas	4. California
5. Florida	6. Oregon	7. Washington	8. Utah
9. Idaho	10. Arizona	11. New Mexico	12. Nevada
13. Montana	14. Wyoming	15. Colorado	16. North Dakota
17. South Dakota	18. Nebraska	19. Kansas	20. Oklahoma
21. Arkansas	22. Louisiana	23. Missouri	24. Iowa
25. Minnesota	26. Illinois	27. Tennessee	28. Kentucky
29. Mississippi	30. Wisconsin	31. Indiana	32. Ohio
33. Alabama	34. Georgia	35. South Carolina	36. North Carolina
37. Virginia	38. Maryland	39. Delaware	40. New Jersey
41. Pennsylvania	42. New York	43. Rhode Island	44. Connecticut
45. Massachusetts	46. Vermont	47. New Hampshire	48. Maine
49. West Virginia	50. Michigan	51. Puerto Rico	

A dot on **Map 3** locates the capital of each state in the United States, as well as Puerto Rico. Place the letter of the capital city next to the corresponding dot on the map.

A. Sacramento	B. Helena	C. Phoenix	D. Salem
E. Carson City	F. Salt Lake City	G. Santa Fe	H. Denver
I. Cheyenne	J. Boise	K. Olympia	L. Austin
M. Oklahoma City	N. Topeka	O. Lincoln	P. Pierre
Q. Bismarck	R. St. Paul	S. Des Moines	T. Jefferson City
U. Little Rock	V. Baton Rouge	W. Madison	X. Springfield
Y. Indianapolis	Z. Frankfort	a. Nashville	b. Jackson
c. Montgomery	d. Tallahassee	e. Atlanta	f. Columbia
g. Raleigh	h. Richmond	i. Charleston	j. Annapolis
k. Dover	l. Harrisburg	m. Trenton	n. Albany
o. Hartford	p. Providence	q. Boston	r. Concord
s. Montpelier	t. Augusta	u. Honolulu	v. Juneau
w. San Juan	x. Columbus	y. Lansing	

Unit 1: Political Geography of Western Hemisphere Nations

C. United States and Capitals (cont.)

MAP 3

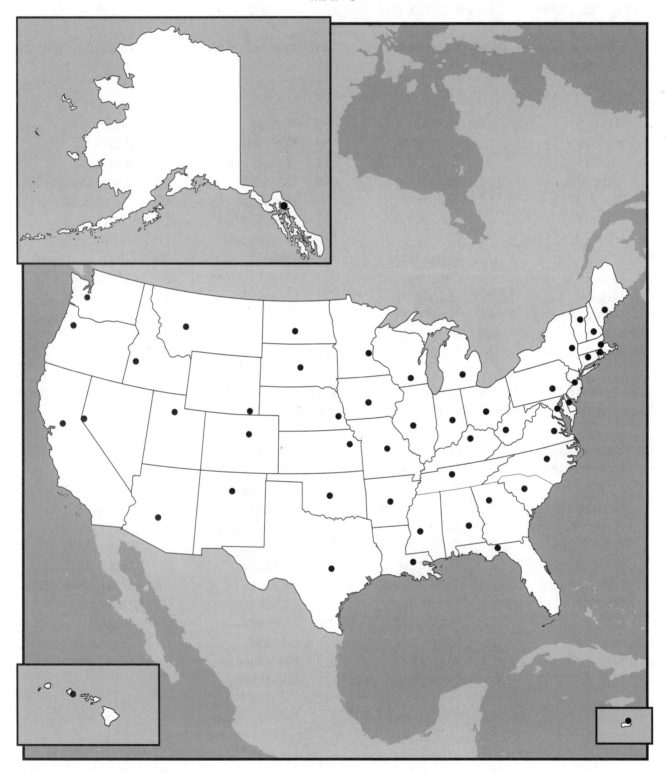

Name: _____ Date: _____

Unit 1: Political Geography of Western Hemisphere Nations

C. United States and Capitals—Pretest Practice

Albany	Annapolis	Atlanta	Augusta	Austin
Baton Rouge	Bismarck	Boise	Boston	Carson City
Charleston	Cheyenne	Columbia	Columbus	Concord
Des Moines	Denver	Dover	Frankfort	Harrisburg
Hartford	Helena	Honolulu	Indianapolis	Jackson
Jefferson City	Juneau	Lansing	Lincoln	Little Rock
Madison	Montgomery	Montpelier	Nashville	Oklahoma City
Olympia	Phoenix	Pierre	Providence	Raleigh
Richmond	Sacramento	Salem	Salt Lake City	San Juan
Santa Fe	Springfield	St. Paul	Tallahassee	Topeka
Trenton				

Use the names of the capital cities above to complete the following.

1. The capital of Alaska is _____.

2. The capital of Hawaii is _____.

3. The capital of Texas is _____.

4. The capital of California is _____.

5. The capital of Florida is _____.

6. The capital of Oregon is _____.

7. The capital of Washington is _____.

8. The capital of Utah is _____.

9. The capital of Idaho is _____.

10. The capital of Arizona is _____.

11. The capital of New Mexico is _____.

12. The capital of Nevada is _____.

13. The capital of Montana is _____.

14. The capital of Wyoming is _____.

15. The capital of Colorado is _____.

16. The capital of North Dakota is _____.

17. The capital of South Dakota is _____.

18. The capital of Nebraska is _____.

19. The capital of Kansas is _____.

Name: _____ Date: _____

Unit 1: Political Geography of Western Hemisphere Nations

C. United States and Capitals—Pretest Practice (cont.)

20. The capital of Oklahoma is _____.
21. The capital of Arkansas is _____.
22. The capital of Louisiana is _____.
23. The capital of Missouri is _____.
24. The capital of Iowa is _____.
25. The capital of Minnesota is _____.
26. The capital of Illinois is _____.
27. The capital of Tennessee is _____.
28. The capital of Kentucky is _____.
29. The capital of Mississippi is _____.
30. The capital of Wisconsin is _____.
31. The capital of Indiana is _____.
32. The capital of Ohio is _____.
33. The capital of Alabama is _____.
34. The capital of Georgia is _____.
35. The capital of South Carolina is _____.
36. The capital of North Carolina is _____.
37. The capital of Virginia is _____.
38. The capital of Maryland is _____.
39. The capital of Delaware is _____.
40. The capital of New Jersey is _____.
41. The capital of Pennsylvania is _____.
42. The capital of New York is _____.
43. The capital of Rhode Island is _____.
44. The capital of Connecticut is _____.
45. The capital of Massachusetts is _____.
46. The capital of Vermont is _____.
47. The capital of New Hampshire is _____.
48. The capital of Maine is _____.
49. The capital of West Virginia is _____.
50. The capital of Michigan is _____.
51. The capital of Puerto Rico is _____.

Unit 1: Political Geography of Western Hemisphere Nations

C. United States and Capitals—Test

Circle the letter of the correct answer.

1. The capital of Wisconsin is
    a) Springfield   b) Madison   c) Lansing   d) Olympia.

2. The capital of Missouri is
    a) Jefferson City   b) Carson City   c) Topeka   d) Little Rock.

3. The capital of Massachusetts is
    a) Salem   b) Concord   c) Augusta   d) Boston.

4. The capital of Tennessee is
    a) Frankfort   b) Nashville   c) Memphis   d) Columbus.

5. The capital of Mississippi is
    a) Baton Rouge   b) Tallahassee   c) Montgomery   d) Jackson.

6. The capital of Alaska is
    a) Sacramento   b) Augusta   c) Juneau   d) Olympia.

7. The capital of Illinois is
    a) Salem   b) Springfield   c) Lincoln   d) Chicago.

8. The capital of Florida is
    a) Tallahassee   b) Salt Lake City   c) Phoenix   d) Boise.

9. The capital of Oregon is
    a) Santa Fe   b) Tallahassee   c) Salem   d) Boise.

10. The capital of Washington is
    a) Trenton   b) Dover   c) Springfield   d) Olympia.

11. The capital of Utah is
    a) Sacramento   b) Indianapolis   c) Salt Lake City   d) Olympia.

12. The capital of Idaho is
    a) Santa Fe   b) Boise   c) Denver   d) Cheyenne.

13. The capital of Arizona is
    a) Phoenix   b) Salem   c) Boise   d) Santa Fe.

14. The capital of New Mexico is
    a) Phoenix   b) Salem   c) Boise   d) Santa Fe.

15. The capital of Georgia is
    a) Jackson   b) Atlanta   c) Montgomery   d) Tallahassee.

16. The capital of South Carolina is
    a) Columbia   b) Atlanta   c) New Orleans   d) Phoenix.

Name: _____ Date: _____

Unit 1: Political Geography of Western Hemisphere Nations

C. United States and Capitals—Test (cont.)

17. The capital of North Carolina is
    a) Richmond   b) Dover   c) Raleigh   d) Columbia.

18. The capital of Virginia is
    a) Raleigh   b) Columbia   c) Hartford   d) Richmond.

19. The capital of Maryland is
    a) Dover   b) Annapolis   c) Providence   d) Atlanta.

20. The capital of Delaware is
    a) Trenton   b) Harrisburg   c) Raleigh   d) Dover.

21. The capital of New Jersey is
    a) Trenton   b) Dover   c) Richmond   d) Hartford.

22. The capital of Pennsylvania is
    a) Annapolis   b) Harrisburg   c) Richmond   d) Columbia.

23. The capital of New York is
    a) Albany   b) Boston   c) Providence   d) Hartford.

24. The capital of Rhode Island is
    a) Albany   b) Trenton   c) Providence   d) Dover.

25. The capital of Connecticut is
    a) Providence   b) Lincoln   c) Harrisburg   d) Hartford.

Name: _____   Date: _____

Unit 1: Political Geography of Western Hemisphere Nations

D. Central American Countries and Capitals

Using **Map 4** and **an atlas**, complete the following about the Central American countries. Place the number by the correct country.

1. Costa Rica   2. Honduras   3. Mexico   4. Belize   5. Panama
6. Nicaragua   7. El Salvador   8. Guatemala

A dot on **Map 4** locates the capital of each Central American country. Place the letter of the capital city next to the corresponding dot on the map.

a. Panamá   b. Belmopan   c. San Salvador   d. Mexico City
e. Managua   f. Guatemala   g. Tegucigalpa   h. San José

MAP 4

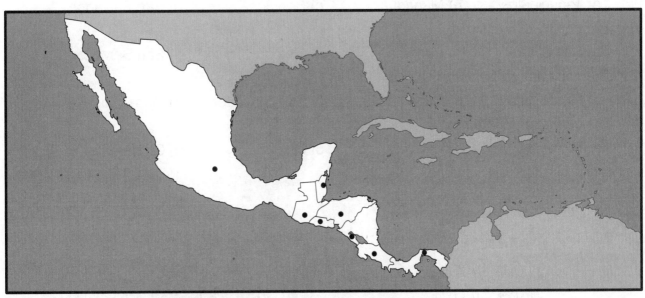

D. Central American Countries and Capitals—Pretest Practice

Use the above capital cities to complete the following.

1. The capital of Belize is _____.
2. The capital of Costa Rica is _____.
3. The capital of Panama is _____.
4. The capital of Honduras is _____.
5. The capital of Nicaragua is _____.
6. The capital of Mexico is _____.
7. The capital of El Salvador is _____.
8. The capital of Guatemala is _____.

Name: _____ Date: _____

Unit 1: Political Geography of Western Hemisphere Nations

D. Central American Countries and Capitals—Test

Circle the letter of the correct answer.

1. The capital of Belize is
   a) San José   b) Panamá   c) Belmopan   d) San Salvador.

2. The capital of Costa Rica is
   a) San José   b) Panamá   c) Belmopan   d) San Salvador.

3. The capital of Panama is
   a) Guatemala   b) Panamá   c) Mexico City   d) Managua.

4. The capital of Honduras is
   a) Tegucigalpa   b) Managua   c) San José   d) San Salvador.

5. The capital of Nicaragua is
   a) Tegucigalpa   b) Managua   c) San José   d) San Salvador.

6. The capital of Mexico is
   a) San José   b) Panamá   c) Mexico City   d) San Salvador.

7. The capital of El Salvador is
   a) Guatemala   b) Panamá   c) San Salvador   d) Managua.

8. The capital of Guatemala is
   a) Managua   b) Panamá   c) Mexico City   d) Guatemala.

Name: _____ Date: _____

Unit 1: Political Geography of Western Hemisphere Nations

E. West Indies (Caribbean) Countries and Capitals

Using **Map 5** and **an atlas**, complete the following about the West Indies (Caribbean). Place the number by the corresponding country.

1. Cuba   2. Haiti   3. Dominican Republic   4. Jamaica
5. Puerto Rico   6. Trinidad and Tobago

Dots on **Map 5** locate the capitals of the West Indies (Caribbean) countries. Place the letter of the city by the corresponding dot on the map.

a. Havana   b. Kingston   c. Port-au-Prince   d. Santo Domingo
e. San Juan   f. Port of Spain

MAP 5

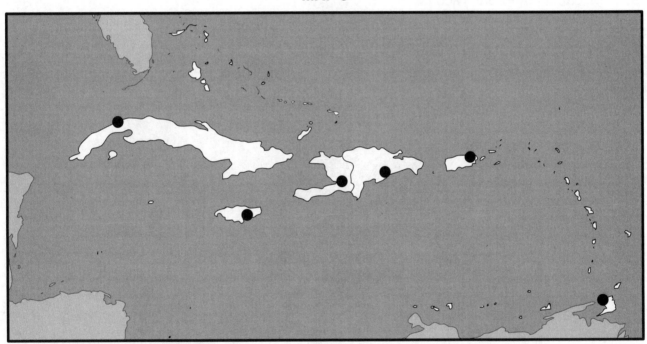

E. West Indies (Caribbean) Countries and Capitals—Pretest Practice

Complete each of the following.

1. The capital of Cuba is _____.
2. The capital of Haiti is _____.
3. The capital of Jamaica is _____.
4. The capital of the Dominican Republic is _____.
5. The capital of Puerto Rico is _____.
6. The capital of Trinidad and Tobago is _____.

Name: _____ Date: _____

Unit 1: Political Geography of Western Hemisphere Nations

E. West Indies (Caribbean) Countries and Capitals—Test

Circle the letter of the correct answer.

1. The capital of Cuba is
   a) Santo Domingo   b) Port of Spain   c) Havana   d) Kingston.

2. The capital of Haiti is
   a) Port-au-Prince   b) Havana   c) Kingston   d) Santo Domingo.

3. The capital of Jamaica is
   a) Santo Domingo   b) San Juan   c) Havana   d) Kingston.

4. The capital of the Dominican Republic is
   a) Santo Domingo   b) San Juan   c) Port-au-Prince   d) Kingston.

5. The capital of Puerto Rico is
   a) Port of Spain   b) San Juan   c) Port-au-Prince   d) Kingston.

6. The capital of Trinidad and Tobago is
   a) Port of Spain   b) San Juan   c) Port-au-Prince   d) Kingston.

Name: _____ Date: _____

Unit 1: Political Geography of Western Hemisphere Nations

F. South American Countries and Capitals

Using **Map 6** and **an atlas**, complete the following about the South American countries. Place the number by the corresponding country.

1. Colombia	2. Argentina	3. Venezuela	4. Chile
5. Ecuador	6. Guyana	7. Bolivia	8. Peru
9. Brazil	10. Paraguay	11. Suriname	12. French Guiana
13. Uruguay			

Dots on **Map 6** locate the capitals of the South American countries. Place the letter of the capital city next to the corresponding dot on the map.

a. Bogotá	b. Caracas	c. Quito	d. Lima
e. Georgetown	f. Cayenne	g. Paramaribo	h. Montevideo
i. Buenos Aires	j. Asunción	k. Brasília	l. Santiago
m. La Paz			

F. South American Countries and Capitals—Pretest Practice

Use the names of the capital cities above to complete the following.

1. The capital of Venezuela is _____.

2. The capital of Chile is _____.

3. The capital of Colombia is _____.

4. The capital of Ecuador is _____.

5. The capital of Brazil is _____.

6. The capital of Uruguay is _____.

7. The capital of Paraguay is _____.

8. The capital of Peru is _____.

9. The capital of Bolivia is _____.

10. The capital of Suriname is _____.

11. The capital of French Guiana is _____.

12. The capital of Guyana is _____.

13. The capital of Argentina is _____.

Name: _____ Date: _____

Unit 1: Political Geography of Western Hemisphere Nations

F. South American Countries and Capitals (cont.)

MAP 6

Name: _____ Date: _____

Unit 1: Political Geography of Western Hemisphere Nations

F. South American Countries and Capitals—Test

1. The capital of Venezuela is
    a) Bogotá   b) Caracas   c) Georgetown   d) Santiago.
2. The capital of Chile is
    a) Bogotá   b) Caracas   c) Georgetown   d) Santiago.
3. The capital of Colombia is
    a) Bogotá   b) Caracas   c) Georgetown   d) Santiago.
4. The capital of Ecuador is
    a) Caracas   b) Lima   c) Brasília   d) Quito.
5. The capital of Brazil is
    a) Lima   b) Caracas   c) Brasília   d) Santiago.
6. The capital of Uruguay is
    a) Montevideo   b) Asunción   c) Buenos Aires   d) Quito.
7. The capital of Paraguay is
    a) Montevideo   b) Asunción   c) La Paz   d) Bogotá.
8. The capital of Peru is
    a) Paramaribo   b) Lima   c) Cayenne   d) Quito.
9. The capital of Bolivia is
    a) Cayenne   b) Asunción   c) La Paz   d) Montevideo.
10. The capital of Suriname is
    a) Paramaribo   b) Lima   c) Cayenne   d) Quito.
11. The capital of French Guiana is
    a) Buenos Aires   b) Paramaribo   c) Cayenne   d) Lima.
12. The capital of Guyana is
    a) Paramaribo   b) Lima   c) Cayenne   d) Georgetown.
13. The capital of Argentina is
    a) Paramaribo   b) Buenos Aires   c) Montevideo   d) Georgetown.

Name: _____ Date: _____

Unit 1: Political Geography of Western Hemisphere Nations

G. Identifying and Locating Countries

For each country on the list below, there is an outline shown. Use a map to identify the country and write the name of the country from the list on blank A. On blank B, write the location as **North America, South America, Central America, or West Indies (Caribbean Islands).**

| **Argentina** | **Brazil** | **Canada** | **Chile** | **Colombia** | **Guatemala** |
| **Mexico** | **Panama** | **Paraguay** | **Peru** | **Puerto Rico** | **United States** |

1.

A. _____

B. _____

2.

A. _____

B. _____

3.

A. _____

B. _____

4.

A. _____

B. _____

5.

A. _____

B. _____

6.

A. _____

B. _____

7.

A. _____

B. _____

8.

A. _____

B. _____

9.

A. _____

B. _____

10.

A. _____

B. _____

11.

A. _____

B. _____

12.

A. _____

B. _____

Name: _____ Date: _____

Unit 2: Population and Area of Western Hemisphere Nations

Each country below has the area listed to the nearest 1,000 or 10,000 square miles. The population is listed to the nearest 100,000. Complete the chart using the area and population figures for each country. (Rank area and population on chart from 1–22, highest to lowest.)

A. Area		B. Population	
Argentina	1,070,000	Argentina	38,000,000
Brazil	3,290,000	Brazil	175,000,000
Canada	3,850,000	Canada	32,000,000
Chile	290,000	Chile	16,000,000
Colombia	440,000	Colombia	41,000,000
Costa Rica	20,000	Costa Rica	4,000,000
Cuba	43,000	Cuba	11,000,000
Dominican Republic	19,000	Dominican Republic	8,500,000
Ecuador	109,000	Ecuador	13,500,000
El Salvador	8,000	El Salvador	6,500,000
Guyana	83,000	Guyana	700,000
Haiti	11,000	Haiti	7,000,000
Honduras	43,000	Honduras	6,500,000
Jamaica	4,000	Jamaica	2,700,000
Mexico	760,000	Mexico	102,000,000
Panama	30,000	Panama	2,900,000
Paraguay	160,000	Paraguay	5,800,000
Peru	500,000	Peru	27,500,000
Trinidad and Tobago	2,000	Trinidad and Tobago	1,200,000
United States	3,620,000	United States	280,000,000
Uruguay	68,000	Uruguay	3,400,000
Venezuela	350,000	Venezuela	24,000,000

Name: _____   Date: _____

Unit 2: Population and Area of Western Hemisphere Nations

A./B. Area and Population (cont.)

Country	Area (Sq. Miles)	Rank	Population	Rank
1. Argentina				
2. Brazil				
3. Canada				
4. Chile				
5. Colombia				
6. Costa Rica				
7. Cuba				
8. Dominican Republic				
9. Ecuador				
10. El Salvador				
11. Guyana				
12. Haiti				
13. Honduras				
14. Jamaica				
15. Mexico				
16. Panama				
17. Paraguay				
18. Peru				
19. Trinidad and Tobago				
20. United States				
21. Uruguay				
22. Venezuela				

Name: _____   Date: _____

Unit 2: Population and Area of Western Hemisphere Nations

C. Comparing the Areas of Countries

On the blanks below, list the names of the ten largest countries in **area** (square miles) from largest to smallest from the chart on the previous page.

1. _____   2. _____   3. _____

4. _____   5. _____   6. _____

7. _____   8. _____   9. _____

10. _____

Write the name of each of the above countries on the correct blanks below to show the location.

North America: _____

Central America: _____

South America: _____

West Indies (Caribbean): _____

D. Comparing the Populations of Countries

On the blanks below, list the names of the ten largest countries in **population** (number of people) from largest to smallest from the chart on the previous page.

1. _____   2. _____   3. _____

4. _____   5. _____   6. _____

7. _____   8. _____   9. _____

10. _____

Write the name of each of the above countries on the correct blanks below to show the location.

North America: _____

Central America: _____

South America: _____

West Indies (Caribbean): _____

Name: _____ Date: _____

Unit 2: Population and Area of Western Hemisphere Nations

D. Comparing the Populations of Countries—Pie Graphs

Countries are often compared on the basis of population. However, it is important to also ask other questions about the population, like the age of the population, religious practices, and the languages spoken, so we can better understand how one nation may differ from another. Many times, pie graphs are used to compare the populations of countries. Pie graphs are made by dividing a circle to show the information. The circle represents 100% of what is being compared. Since a circle is 360°, the circle can be divided to show various percentages.

Example: Country A has 200 people, Country B has 300 people, and Country C has 500 people. Country A + Country B + Country C = 1,000 people. To determine the percentages of each: 300 ÷ 1,000 = 0.30 x 100 = **30%**; 200 ÷ 1,000 = 0.20 x 100 = **20%**; 500 ÷ 1,000 = 0.50 x 100 = **50%**. **20% + 30% + 50% = 100%.** To determine the part of the pie graph for each country, multiply **0.20 x 360° = 72°; 0.30 x 360 = 108°; 0.50 x 360° = 180°; 72° + 108° + 180° = 360°.**

Pie Graph A

Populations of Countries A, B, and C

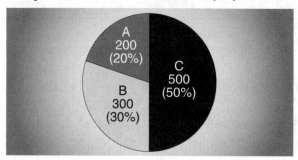

Using **Pie Graph A** above, circle the letter of the correct answer.

1. Pie graph A compares the   a) age of people in the countries   b) population in the countries   c) religion of the countries.

2. The smallest country in population is   a) Country B   b) Country C   c) Country A.

3. The largest country in population is   a) Country B   b) Country C   c) Country A.

4. Country C has   a) more people than Countries A and B together   b) fewer people than Countries A and B together   c) the same number of people as Countries A and B together.

5. The total number of people shown on the pie chart is a) 200   b) 300   c) 500   d) 1000.

6. The complete pie chart represents   a) 100%   b) 20%   c) 30%   d) 50%.

7. Country A has a) 20% b) 30% c) 50% of the total number of people shown on the pie graph.

8. Country B has a) 20% b) 30% c) 50% of the total number of people shown on the pie graph.

9. Country C has a) 20% b) 30% c) 50% of the total number of people shown on the pie graph.

Name: _____ Date: _____

Unit 2: Population and Area of Western Hemisphere Nations

D. Comparing the Populations of Countries—Pie Graphs (cont.)

Pie Graph B

Populations of Three Western Hemisphere Nations

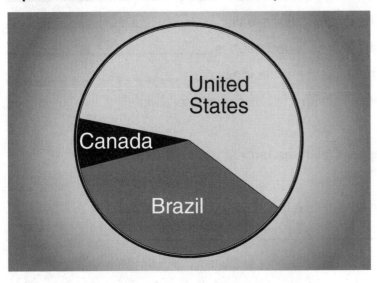

Using **Pie Graph B,** circle the letter of the correct answer.

1. The country with the largest population is   a) Brazil   b) the United States   c) Canada.

2. The country with the smallest population is a) Brazil   b) the United States   c) Canada.

3. The population of Brazil is approximately   a) one time   b) ten times   c) five times

    the size of Canada.

4. The population of the United States is approximately   a) five times   b) one and one-half

    times   c) ten times   greater than the population of Brazil.

5. The population of the United States is approximately   a) three times   b) twelve times

    c) eight and one-half times   greater than the population of Canada.

6. The population of Canada and Brazil together is   a) less than   b) greater than

    c) the same as   the population of the United States.

Name: _____ Date: _____

Unit 2: Population and Area of Western Hemisphere Nations

E. Comparing the Populations and Areas of Countries—Pretest Practice

Complete each of the following.

1. The South American country with the largest land area is _____.

2. The North American country with the largest land area is _____.

3. The South American country with the largest population is _____.

4. The North American country with the largest population is _____.

5. The country in the Western Hemisphere with the largest land area is

    _____.

6. The country in the Western Hemisphere with the largest population is

    _____.

7. The three countries in South America with the largest land areas are

    _____, _____, and

    _____.

8. The three countries in South America with the largest populations are

    _____, _____, and _____.

9. The three countries in the Western Hemisphere with the largest land areas are

    _____, _____, and

    _____.

10. The three countries in the Western Hemisphere with the largest populations are

    _____, _____, and

    _____.

11. The country in Central America with the largest land area

    is _____.

12. The two countries in Central America with the largest

    populations are _____

    and _____.

13. The country in the West Indies (Caribbean) with the larg-

    est land area is _____.

14. The country in the West Indies (Caribbean) with the larg-

    est population is _____.

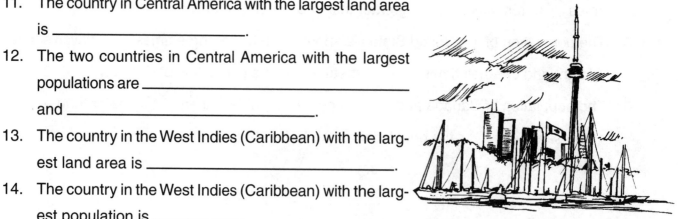

Name: _____ Date: _____

Unit 2: Population and Area of Western Hemisphere Nations

E. Comparing the Populations and Areas of Countries—Test

Circle the letter of the correct answer.

1. The South American country with the largest land area is
   a) Uruguay   b) Brazil   c) Argentina   d) Chile.

2. The North American country with the largest land area is
   a) Mexico   b) the United States   c) Canada.

3. The South American country with the largest population is
   a) Colombia   b) Venezuela   c) Paraguay   d) Brazil.

4. The North American country with the largest population is
   a) the United States   b) Canada   c) Mexico.

5. The country in the Western Hemisphere with the largest land area is
   a) Canada   b) the United States   c) Brazil   d) Mexico.

6. The country in the Western Hemisphere with the largest population is
   a) Canada   b) the United States   c) Brazil   d) Mexico.

7. The three countries in South America with the largest land areas are
   a) Brazil, Venezuela, and Chile   b) Peru, Argentina, and Brazil
   c) Paraguay, Uruguay, and Bolivia.

8. The three countries in South America with the largest populations are
   a) Brazil, Venezuela, and Chile   b) Colombia, Argentina, and Brazil
   c) Paraguay, Uruguay, and Bolivia.

9. The three countries in the Western Hemisphere with the largest land areas are
   a) the United States, Canada, and Brazil   b) the United States, Brazil, and Mexico
   c) the United States, Canada, and Venezuela.

10. The three countries in the Western Hemisphere with the largest populations are
    a) the United States, Canada, and Brazil   b) the United States, Brazil, and Mexico
    c) the United States, Canada, and Venezuela.

11. The country in Central America with the largest land area is
    a) Honduras   b) El Salvador   c) Costa Rica   d) Panama.

12. The countries in Central America with the largest populations are
    a) Honduras and Belize   b) El Salvador and Honduras   c) Costa Rica and Panama
    d) Panama and El Salvador.

13. The country in the West Indies (Caribbean) with the largest land area is
    a) Trinidad   b) Haiti   c) Cuba   d) Dominican Republic.

14. The country in the West Indies (Caribbean) with the largest population is
    a) Trinidad   b) Haiti   c) Cuba   d) Dominican Republic.

Name: _____ Date: _____

Unit 3: Physical Features of the Western Hemisphere

A. Rivers

On **Map 7**, each of the following rivers is identified with a letter. Using **Map 7** and **an atlas**, write the name of the river the letter locates on the blank beside that letter.

Amazon	**Colorado**	**Columbia**	**Fraser**	**Hudson**	**Illinois**
Mackenzie	**Magdalena**	**Mississippi**	**Missouri**	**Ohio**	**Orinoco**
Parana	**Platte**	**Potomac**	**Rio Grande**	**Sacramento**	
San Joaquin	**Snake**	**St. Lawrence**	**Tennessee**	**Yukon**	

a. _____ b. _____ c. _____

d. _____ e. _____ f. _____

g. _____ h. _____ i. _____

j. _____ k. _____ l. _____

m. _____ n. _____ o. _____

p. _____ q. _____ r. _____

s. _____ t. _____ u. _____

v. _____

MAP 7

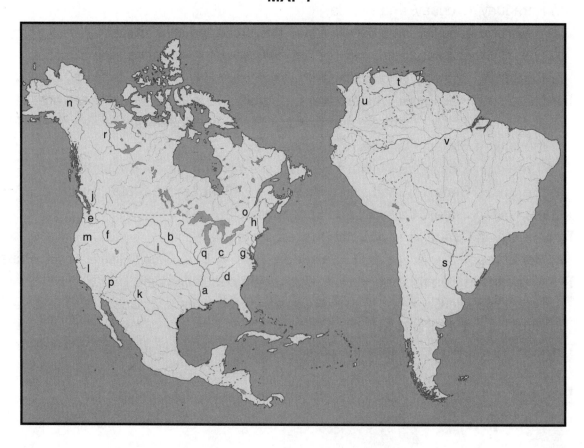

Name: _____ Date: _____

Unit 3: Physical Features of the Western Hemisphere

A. Rivers—Crossword Puzzle

Use the rivers from page 28 to complete the crossword puzzle.

ACROSS

2. A large river that empties into the Gulf of Mexico at New Orleans
9. A river that forms the boundary between Washington and Oregon and flows into the Pacific Ocean
11. A river that flows south in New York to New York City
13. A river in Venezuela
14. A river that flows north through Tennessee and Kentucky and into the Ohio River at Paducah, Kentucky
15. A river that flows into the Missouri
16. A river that begins in Colorado and flows through the Grand Canyon and into the Gulf of California
19. A river that flows north into the Northwest Territories, Canada

DOWN

1. River that forms the boundary between the United States and Mexico (two words)
3. River that flows from Lake Ontario to the Gulf of St. Lawrence
4. A river that flows through Idaho and Oregon and into the Columbia River
5. Washington, D.C., is located on this river
6. A river in British Columbia
7. A famous river in Alaska
8. A river in Brazil near the equator
10. A river that forms the boundary between Ohio and Kentucky and Indiana and Kentucky and flows into the Mississippi River south of St. Louis
12. A river that flows from Montana through North Dakota, South Dakota, along the border between Nebraska and Iowa, and into the Mississippi River near St. Louis
17. A river in California (two words)
18. A river in Argentina that flows into the Rio de la Plata

Name: _____　　Date: _____

Unit 3: Physical Features of the Western Hemisphere

B. Mountains and Volcanoes

1. The mountains below are identified on **Map 8** with upside-down "V's" and semicircles and a letter. Using an atlas, write the name of the mountains indicated by each letter on the blank by the letter.

Adirondacks　　**Alaskan Range**　　**Andes**　　**Appalachians**　　**Blue Ridge**
Brooks Range　　**Cascades**　　**Catskills**　　**Rockies**　　**Sierra Nevadas**
Brazilian Highlands

a. _____　　b. _____　　c. _____

d. _____　　e. _____　　f. _____

g. _____　　h. _____　　i. _____

j. _____　　k. _____

2. Listed below are the names of high mountain peaks in North America, Central America, and South America. The height of the peak to the nearest 100 feet is listed in parentheses. Many of these peaks are volcanic peaks.

Mt. Aconcagua (22,800)　　Mt. Chimborazo (20,700)　　Mt. Cotopaxi (19,300)
Mt. Iztaccíhuatl (17,200)　　Mt. Logan (19,600)　　Mt. McKinley (20,300)
Mt. Popocatépetl (17,900)　　Mt. Rainier (14,400)　　Mt. Shasta (14,200)
Mt. Whitney (14,500)　　Pikes Peak (14,100)

The mountain peaks are identified on **Map 8** with the symbol $\wedge$ and a letter. Locate each peak on Map 8 and, using an atlas, write the name on the blank below with the corresponding letter.

a. _____　　b. _____　　c. _____

d. _____　　e. _____　　f. _____

g. _____　　h. _____　　i. _____

j. _____　　k. _____

Unit 3: Physical Features of the Western Hemisphere

B. Mountains and Volcanoes (cont.)

MAP 8

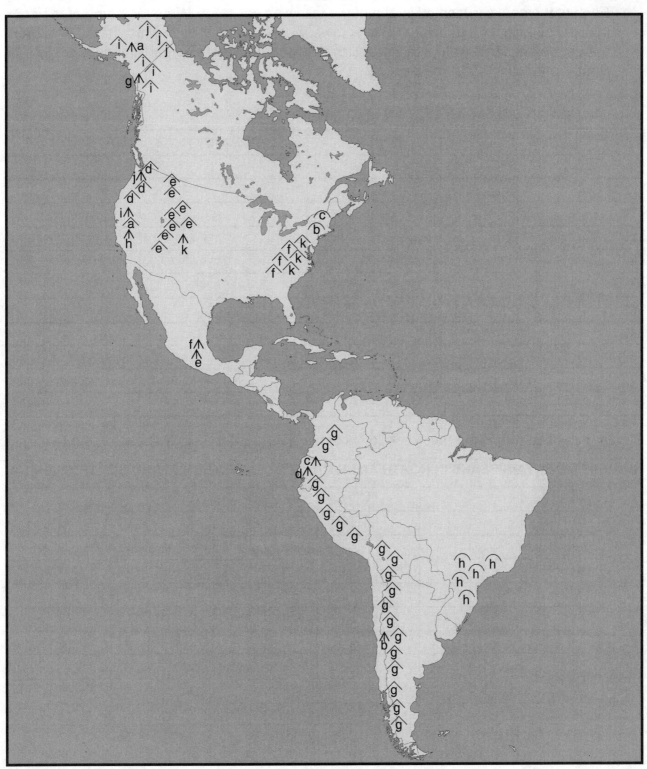

Name: _____ Date: _____

Unit 3: Physical Features of the Western Hemisphere

B. Mountains and Volcanoes (cont.)

3. Using the previous exercise, arrange the mountain peaks' heights from highest to lowest. Each of the peaks is located in one of the following countries. Write the name of the peak in order by height (from a, highest to k, lowest), and the country in which it is located from the list below. Then write "NA" for North America, "CA" for Central America, or "SA" for South America to show the location.

Argentina   Canada   Ecuador   Mexico   United States

Peak	Height	Country	NA, CA, or SA
a.			
b.			
c.			
d.			
e.			
f.			
g.			
h.			
i.			
j.			
k.			

Shade in the squares in the bar graph to show the approximate height of each mountain.

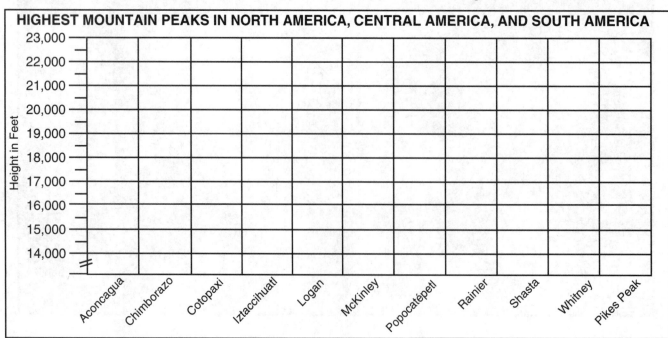

Name: _____ Date: _____

Unit 3: Physical Features of the Western Hemisphere

C. Plains and Valleys

The plains and valleys below are located on **Map 9**. A number is used to locate each feature. Using an atlas, circle the letter for the correct country/state in which it is located.

1. Pampas: a) Brazil   b) Argentina   c) Chile
2. Great Plains: a) Peru   b) United States   c) Mexico
3. Orinoco Plain: a) Uruguay   b) Bolivia   c) Venezuela
4. Gran Chaco: a) Paraguay   b) Venezuela   c) Cuba
5. Great Central Valley: a) California   b) Chile   c) Canada
6. Vale of Chile: a) California   b) Chile   c) Canada
7. Atlantic Coastal Plain: a) Chile   b) United States   c) Mexico

D. Plateaus

The plateaus below are located by a number under an arc on **Map 9**. Using an atlas, circle the letter of the correct answer that indicates the location of the plateau.

1. Altiplano: a) Bolivia   b) Haiti   c) Saskatchewan
2. Edwards: a) British Columbia   b) Panama   c) Texas
3. Ozark: a) Missouri   b) Chile   c) Argentina
4. Patagonia: a) Ontario   b) Argentina c) Brazil
5. Cumberland: a) Costa Rica   b) Canada   c) Tennessee

Unit 3: Physical Features of the Western Hemisphere

C. Plains and Valleys/D. Plateaus (cont.)

MAP 9

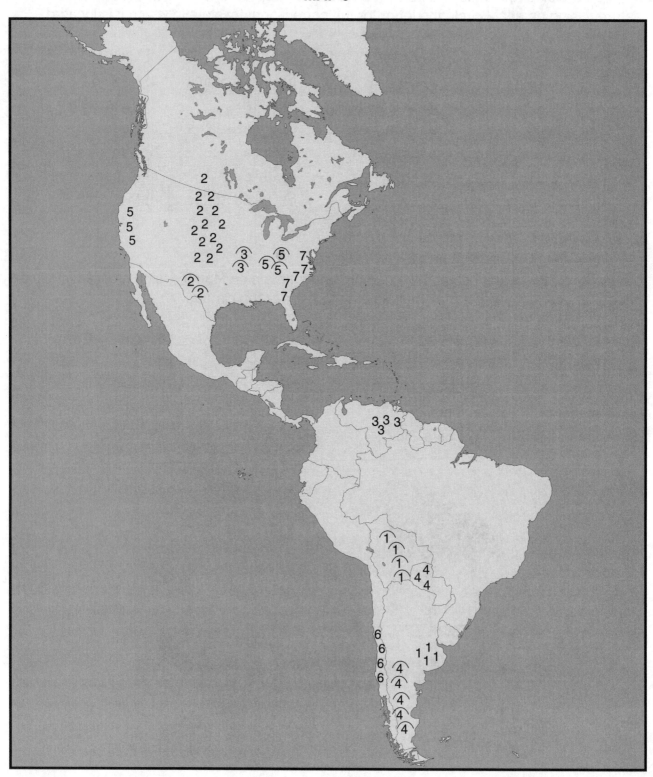

Name: _____ Date: _____

Unit 3: Physical Features of the Western Hemisphere

E. Lakes, Seas, Gulfs, and Bays

On **Map 10**, numbers are used to locate the lakes listed below. Using an atlas, circle the answer that indicates the location(s) of the lake.

1.	Lake Titicaca:	a) Chile	b) Bolivia and Peru	c) Canada
2.	Great Salt Lake:	a) Brazil	b) Utah	c) Florida
3.	Lake Superior:	a) U.S. and Canada	b) Alaska	c) Uruguay
4.	Great Bear Lake:	a) Montana	b) Honduras	c) Canada
5.	Lake Erie:	a) California	b) U.S. and Canada	c) Yukon
6.	Lake Champlain:	a) New York	b) Florida	c) Peru
7.	Great Slave Lake:	a) Mexico	b) Brazil	c) Canada
8.	Lake Winnipeg:	a) Manitoba	b) Utah	c) Ontario
9.	Lake Okeechobee:	a) Florida	b) Newfoundland	c) Iowa
10.	Lake Managua:	a) Honduras	b) Nicaragua	c) Belize
11.	Lake Poopó:	a) Brazil	b) Chile	c) Bolivia
12.	Lake Maracaibo:	a) Peru	b) Venezuela	c) Brazil
13.	Lake Nicaragua:	a) Honduras	b) Nicaragua	c) Belize

Using **Map 10** and **an atlas**, place the number of the sea, gulf, or bay on the map to show its location.

1. Gulf of Mexico   2. Hudson Bay   3. Caribbean Sea   4. Chesapeake Bay
5. Gulf of California

Name: _____   Date: _____

Unit 3: Physical Features of the Western Hemisphere

E. Lakes, Seas, Gulfs, and Bays (cont.)

MAP 10

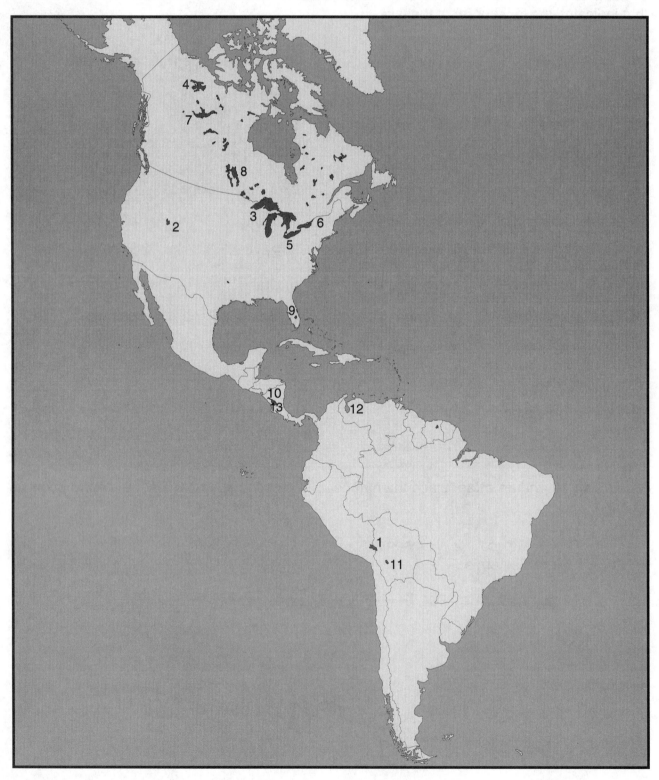

Unit 3: Physical Features of the Western Hemisphere

F. Physical Features—Pretest Practice

Aconcagua	**Altiplano**	**Amazon**	**Argentina**	**Bolivia**	**California**
Caribbean	**Chesapeake**	**Gran Chaco**	**Great Plains**	**Mackenzie**	**McKinley**
Mississippi	**Pampas**	**Peru**	**Rio Grande**		

1. A large river in the United States that flows south and empties into the Gulf of Mexico is the
   _____.

2. A large river in Brazil that flows east into the Atlantic Ocean is the _____.

3. A river that forms the boundary between the United States and Mexico is the
   _____.

4. This river that flows north through the Northwest Territories of Canada is the
   _____.

5. The highest mountain in the Western Hemisphere is Mt. _____.

6. The highest mountain in the Western Hemisphere is located in the country of
   _____.

7. The highest mountain in the United States is Mt. _____.

8. Mt. Whitney, the highest peak in the continental United States, is located in the state of
   _____.

9. A fertile plains region in Argentina noted for farming and ranching is the _____.

10. A high, cool plateau in Bolivia where most people live is the _____.

11. A plains region in the United States found in North Dakota, South Dakota, Kansas, Nebraska, and Oklahoma is the _____.

12. Located in Paraguay is a region noted for quebracho trees, which have very hard wood. Its bark is used to produce tannin for tanning leather. This region is also noted for yerba maté, a tea-like drink. The region is the _____.

13. Lake Titicaca located on the high Altiplano, which was the home of the Inca Indians, is located in _____ and _____.

14. A large bay located in northern Canada is _____ Bay.

15. Cuba, Jamaica, Haiti, and the Dominican Republic are located in this area. The sea is the _____.

16. The Potomac River flows into this bay. Washington, D.C., is located nearby. The bay is the _____.

Name: _____ Date: _____

Unit 3: Physical Features of the Western Hemisphere

F. Physical Features—Test

Circle the letter of the correct answer.

1. A large river in the United States that flows south and empties into the Gulf of Mexico is the
   a) Tennessee   b) Ohio   c) Mississippi   d) Fraser.

2. A large river in Brazil that flows east into the Atlantic Ocean is the
   a) Amazon   b) Parana   c) Magdalena   d) Hudson.

3. A river that forms the boundary between the United States and Mexico is the
   a) St. Lawrence   b) Rio Grande   c) Missouri   d) Yukon.

4. A river that flows north through the Northwest Territories of Canada is the
   a) Yukon   b) Mackenzie   c) St. Lawrence   d) Orinoco.

5. The highest mountain in the Western Hemisphere is
   a) Mt. Aconcagua   b) Mt. Whitney   c) Mt. Cotopaxi   d) Mt. McKinley.

6. The highest mountain in the Western Hemisphere is located in the country of
   a) Brazil   b) United States   c) Honduras   d) Argentina.

7. The highest mountain in the United States is
   a) Mt. Whitney   b) Mt. Rainier   c) Mt. McKinley   d) Mt. Shasta.

8. Mt. Whitney, the highest peak in the continental United States, is located in the state of
   a) Colorado   b) Washington   c) California   d) Oregon.

9. A fertile plains region in Argentina noted for farming and ranching is the
   a) Llanos   b) Great Plains   c) Vale of Chile   d) Pampas.

10. A high, cool plateau in Bolivia where most people live is the
    a) Great Central Valley   b) Altiplano   c) Llanos   d) Ozark.

11. A plains region in the United States found in North Dakota, South Dakota, Kansas, Nebraska, and Oklahoma is the
    a) Great Plains   b) Great Central Valley   c) Vale of Chile   d) Pampas.

12. A region in Paraguay noted for quebracho trees and yerba maté is the
    a) Gran Chaco   b) Pampas   c) Llanos   d) Patagonia.

13. Lake Titicaca on the high Altiplano, which was the home of the Inca Indians, is located in
    a) Peru and Argentina   b) Paraguay and Bolivia   c) Bolivia and Peru
    d) Brazil and Venezuela.

14. A large bay located in northern Canada is the   a) Chesapeake   b) Hudson
    c) California   d) Delaware.

15. Cuba, Jamaica, Haiti, and the Dominican Republic are located in this area. The sea is the
    a) White   b) Arctic   c) Caribbean   d) Red.

16. The Potomac River flows into this bay. Washington, D.C., is located nearby. The bay is the
    a) California   b) St. Lawrence   c) Hudson   d) Chesapeake.

Name: _____ Date: _____

Unit 4: Using Latitude and Longitude

Places on the earth are located using a grid system known as **latitude and longitude**. Use Diagram 1 to complete the following exercise to learn how latitude and longitude are used to determine location.

DIAGRAM 1

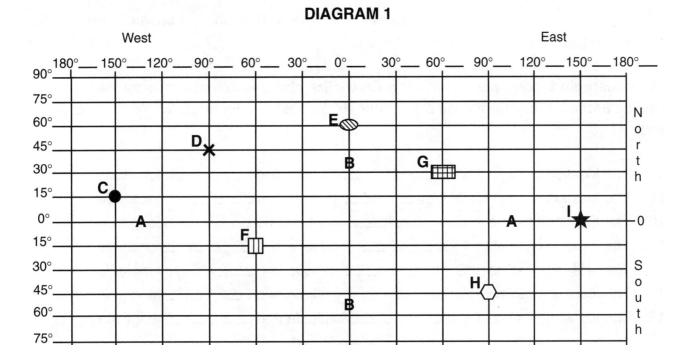

Latitude is measured north and south from the equator. Zero degrees latitude is also known as the **equator**. Latitude locations are always 0 to 90 degrees north or 0 to 90 degrees south. Latitude lines run from east to west but measure distance north or south from the equator.

Zero degrees longitude is also known as the **Prime Meridian**. Longitude locations are always 0 to 180 degrees east or 0 to 180 degrees west. Longitude lines run north and south but measure distance east or west from the Prime Meridian (0 degrees).

1. On the blank next to each number of degrees on Diagram 1, place the letter E (east), W (west), N (north), or S (south).

2. Find "A" on the diagram and write in "equator" along this line. Draw a line over the equator.

3. The equator is   a) 10   b) 50   c) 0   degrees latitude.

4. Latitude is measured   a) east and west   b) north and south   from the equator.

5. Latitude lines run   a) east and west   b) north and south   but measure distance
    a) east and west   b) north and south.

6. Find "B" on the chart and write "Prime Meridian" along the line. Draw a line over the Prime Meridian.

39

Name: _____ Date: _____

Unit 4: Using Latitude and Longitude

7. The Prime Meridian is   a) 10   b) 50   c) 0   degrees longitude.

8. Longitude is measured   a) east and west   b) north and south   from the Prime Meridian.

9. Longitude lines run   a) east and west   b) north and south   but measure distance
   a) east and west   b) north and south.

Using **Diagram 1,** complete the following. On the first blank, write the number for the latitude or longitude location. On the second blank, write "E" for east, "W" for west, "N" for north, or "S" for south.

10. The location of symbol "C" is _____° _____ latitude and _____° _____ longitude.

11. The location of symbol "D" is _____° _____ latitude and _____° _____ longitude.

12. The location of symbol "E" is _____° _____ latitude and _____° _____ longitude.

13. The location of symbol "F" is _____° _____ latitude and _____° _____ longitude.

14. The location of symbol "G" is _____° _____ latitude and _____° _____ longitude.

15. The location of symbol "H" is _____° _____ latitude and _____° _____ longitude.

16. The location of symbol "I" is _____° _____ latitude and _____° _____ longitude.

Using **Map 11** and **an atlas,** circle the letter of the correct answer.

1. A location that is 60 degrees north and 30 degrees west is in
   a) Brazil   b) the north Atlantic Ocean   c) the Pacific Ocean.

2. A location that is 15 degrees south and 60 degrees west is in
   a) Brazil   b) the United States   c) Argentina.

3. A location that is 45 degrees south and 105 degrees west is in
   a) Chile   b) Mexico   c) the Pacific Ocean.

4. A location that is 60 degrees north and 150 degrees west is in
   a) Bolivia   b) Alaska   c) Mexico.

5. A location that is 0 degrees latidude and 75 degrees west is in
   a) Colombia   b) Bolivia   c) Trinidad.

6. A latitude location is given as 30 degrees north. No longitude location is given. The location could be in
   a) Chile or Argentina   b) Mexico or the United States   c) Ecuador or Brazil.

7. A longitude location is given as 75 degrees west. No latitude location is given. The location could be in   a) Chile, Peru, Colombia, Cuba, the United States, or Canada
   b) Brazil, Venezuela, Colombia, Cuba, the United States, or Canada.

Unit 4: Using Latitude and Longitude

8. A location between 15 degrees north and 30 degrees north could be in

  a) Argentina, Bolivia, Paraguay, Peru, Chile, or Brazil

  b) Mexico, Cuba, Puerto Rico, Dominican Republic, or Haiti.

9. Using the list of countries below, place the letter "S" on the blank if the country is located south of the equator. Place the letter "N" on the blank if the country is located north of the equator.

_____ Bolivia   _____ Canada   _____ Chile   _____ Cuba

_____ Mexico   _____ Panama   _____ Peru   _____ United States

_____ Uruguay

MAP 11

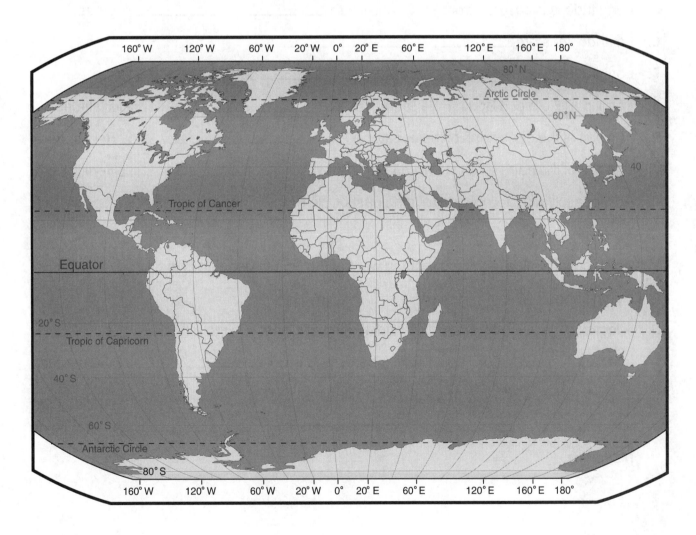

Name: _____ Date: _____

Unit 4: Using Latitude and Longitude—Pretest Practice

Complete the blanks using the following terms. Some terms may be used more than once.

90	180	east	Eastern	equator	latitude
longitude	parallel	Poles	Prime Meridian	north	Northern
south	Southern	west	Western	zero	

1. Latitude is measured north and south of the _____.

2. Longitude is measured east and west from the _____.

3. The line of latitude that is zero degrees latitude is the_____.

4. The line of longitude that is zero degrees longitude is the _____.

5. Longitude is measured from zero degrees to _____ degrees.

6. Latitude lines are _____.

7. Longitude lines meet at the _____.

8. The North Pole and South Pole represent _____ degrees latitude.

9. The half of the earth north of the equator is the _____ Hemisphere.

10. The half of the earth south of the equator is the _____ Hemisphere.

11. Any location with a longitude reading east of the Prime Meridian is in the _____

    Hemisphere.

12. Any location with a longitude reading west of the Prime Meridian is in the _____

    Hemisphere.

13. A location that is 30 degrees north and 45 degrees west is in the _____

    and _____ Hemispheres and is _____ of the equator.

14. A location that is 30 degrees south and 45 degrees east is in the _____

    and _____ Hemispheres and is _____ of the Prime

    Meridian.

Name: _____ Date: _____

Unit 4: Using Latitude and Longitude—Test

Circle the letter of the correct answer.

1. Latitude is measured north and south from the
   a) Prime Meridian   b) equator
   c) International Date Line.

2. Longitude is measured east and west from the
   a) Prime Meridian   b) equator
   c) International Date Line.

3. The line of latitude that is zero degrees latitude is the
   a) Prime Meridian   b) equator
   c) International Date Line.

4. The line of longitude that is zero degrees longitude is the   a) Prime Meridian
   b) equator   c) International Date Line.

5. Longitude is measured from zero degrees to   a) 60   b) 75   c) 100
   d) 180   degrees east or west.

6. Latitude lines   a) run north and south   b) measure distance east and west
   c) are parallel.

7. Longitude lines meet at the   a) equator   b) Prime Meridian   c) Poles
   d) International Date Line.

8. The North Pole and South Pole represent   a) 180   b) 50   c) 90   d) 0   degrees latitude.

9. The half of the earth north of the equator is called the   a) Northern   b) Western
   c) Southern   d) Eastern   Hemisphere.

10. The half of the earth south of the equator is called the   a) Northern   b) Western
    c) Southern   d) Eastern   Hemisphere.

11. Any location with a longitude reading east of the Prime Meridian is in the   a) Northern
    b) Western   c) Southern   d) Eastern   Hemisphere.

12. Any location with a longitude reading west of the Prime Meridian is in the   a) Northern
    b) Western   c) Southern   d) Eastern   Hemisphere.

13. A location that is 30 degrees north and 45 degrees west is in the   a) Eastern
    b) Western   Hemisphere and north of the   c) equator   d) Prime Meridian.

14. A location that is 30 degrees south and 45 degrees east is in the   a) Eastern
    b) Western   Hemisphere and east of the   c) equator   d) Prime Meridian.

Name: _____ Date: _____

Unit 5: Climate in the Western Hemisphere

A. Tundra Climate

The **tundra climate** has long, cold winters with snow on the ground most of the year. Although summers are short, there are 20 to 24 hours of daylight. Even with the long periods of daylight, only one month has an average monthly temperature above 32 degrees. The growing season is too short for trees, so the vegetation consists of moss and lichens. During the long summer days, the soil will thaw on the surface. Because the topmost soil layer thaws, the water from the thawed layer remains on the surface, making it very wet and muddy and a place where many insects can survive. However, below the surface, the soil stays permanently frozen. This permanently frozen layer is known as **permafrost**.

The warmest average monthly temperature is 40°F. The coldest average monthly temperature is -20°F, and there is an average snowfall of six inches. During the long winter, the cold and blowing snow occurs almost daily.

Complete the following.

1. The highest average monthly temperature is _____ degrees.

2. The lowest average monthly temperature is _____ degrees.

3. The difference between the highest average temperature and lowest average temperature is   a) 10°   b) 60°   c) 40°   d) 0°.

Using **Map 12a** and **an atlas**, complete the following.

4. The number 5s indicate the region with a tundra climate. This climate is located in

    a) the United States   b) Argentina   c) Canada   d) Cuba.

5. Place a "+" on the blank if any part of the province or territory has a tundra climate.

    _____ Alberta   _____ Newfoundland   _____ Ontario   _____ Quebec

    _____ Yukon Territory   _____ Nunavut Territory   _____ Northwest Territories

6. The following cities are located on the map with a dot. Place a "+" on the blank if the city is in the tundra climate.

    _____ Montreal   _____ Quebec   _____ Regina   _____ St. John's

    _____ Vancouver   _____ Iqaluit

Name: _____ Date: _____

Unit 5: Climate in the Western Hemisphere

MAP 12a

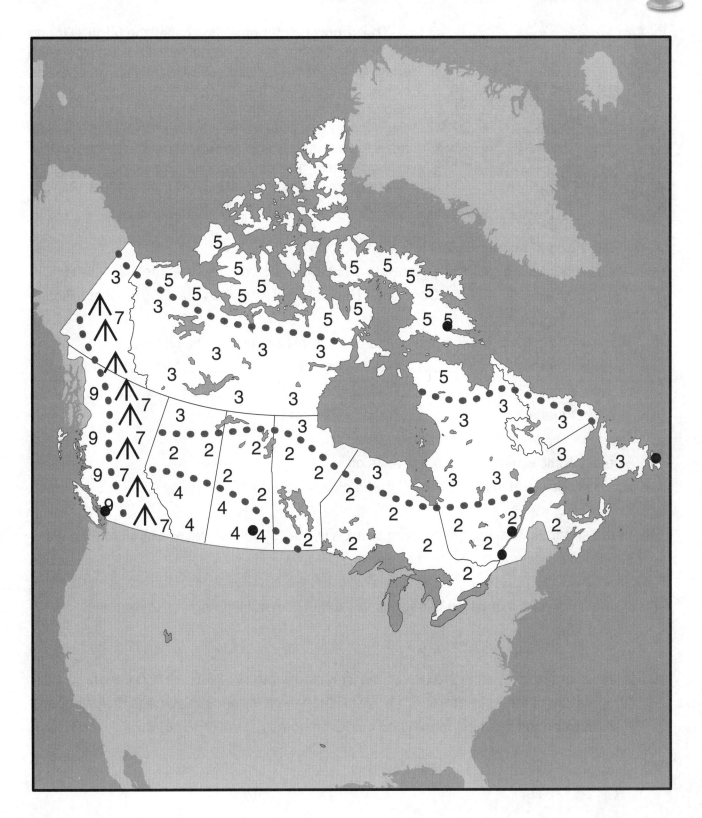

Name: _____ Date: _____

Unit 5: Climate in the Western Hemisphere

B. Subarctic Climate (Taiga)

The **subarctic climate**, often called the **taiga**, is found in Canada. Winters are long and very cold. Summers are short with a growing season of 30 to 60 days. In only one or two months does the average daily temperature rise above 50 degrees Fahrenheit. Snowfall is abundant and remains on the ground for over half the year. This climate is known for large forests of coniferous trees that include pines, firs, and spruces. The forests are often referred to as **boreal forests**. The lumber industry is very important in this climate region.

Using **Map 12a** and **an atlas**, complete the following.

1. The number 3s indicate the region with a subarctic climate. This climate is located in

    a) Venezuela   b) the Dominican Republic   c) Canada   d) Costa Rica.

2. Place a "+" on the blank if the province or territory has a subarctic climate.

    _____ Alberta   _____ Newfoundland   _____ Northwest Territories

    _____ Nova Scotia   _____ Ontario   _____ Prince Edward Island

    _____ Quebec   _____ Nunavut Territory

3. The subarctic climate has large forests of

    a) oak, elm, and maple   b) fir, pine, and spruce.

4. In the subarctic climate, the growing season is   a) 90 to 120 days   b) 60 to 90 days

    c) 30 to 60 days.

5. A very important industry in this region is   a) auto manufacturing   b) farming

    c) lumber.

6. If the average daily temperature of the coldest month is -23°F, and the average daily temperature of the warmest month is +59°F, then the difference in average daily temperature between the warmest month and coldest month is   a) 40°F   b) 82°F   c) 30°.

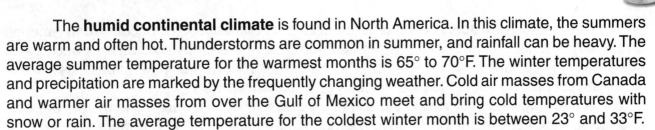

Unit 5: Climate in the Western Hemisphere

C. Humid Continental Climate

The **humid continental climate** is found in North America. In this climate, the summers are warm and often hot. Thunderstorms are common in summer, and rainfall can be heavy. The average summer temperature for the warmest months is 65° to 70°F. The winter temperatures and precipitation are marked by the frequently changing weather. Cold air masses from Canada and warmer air masses from over the Gulf of Mexico meet and bring cold temperatures with snow or rain. The average temperature for the coldest winter month is between 23° and 33°F.

In the humid continental climate, the time between the last killing frost of the spring and the first killing frost of the fall is approximately 180 days. The growing season is the time when crops can grow without a killing frost. Agriculture is very important in this climate, with crops like corn, wheat, and soybeans being grown.

In the humid continental climate, there are many broadleaf trees, such as maple, oak, birch, and poplar.

Using **Maps 12a**, **12b**, and **an atlas**, complete the following.
The number 2s indicate the region with a humid continental climate.

1. The humid continental climate is found in the countries of   a) Mexico and the United States   b) the United States and Canada   c) Canada and Mexico.

2. Place a "+" on the blank to indicate if a humid continental climate is found in any part of Canada.   _____ Alberta   _____ Nova Scotia   _____ Ontario   _____ Quebec   _____ Saskatchewan   _____ Yukon Territory

3. Place a "+" on the blank to indicate if a humid continental climate is found in any part of these states.   _____ California   _____ Florida   _____ Maine   _____ Minnesota   _____ New York   _____ Ohio   _____ Texas   _____ Vermont   _____ Wisconsin

Extremely high summer temperatures and cold winters are found in the humid continental climate. These high and low temperatures occur in North America because as North America extends toward the North Pole, the landmass becomes much larger. In the summer, this large landmass becomes very warm. In the winter, it becomes very cold. In South America, the landmass becomes more narrow and does not extend to as high a latitude as the North American continent. The humid continental climate is not found on the South American continent.

Using **Maps 12b**, **12c**, and **an atlas**, complete the following. (North America and South America)
4. As the North American continent gets closer to the North Pole, the landmass becomes
   a) wider   b) narrower.
5. As the South American continent gets closer to the South Pole, the landmass becomes
   a) wider   b) narrower.
6. True or False: If the landmass of the South American continent became wider and extended to a higher latitude, the humid continental climate would be found in South America.

Name: _____ Date: _____

Unit 5: Climate in the Western Hemisphere

D. Humid Subtropical Climate

The **humid subtropical climate** is found on the southeast coast of both North America and South America. In North America, this is the region where crops like cotton, tobacco, and rice are very important. In South America, this is the region known as the Pampas. It is famous for large cattle ranches and the production of grain crops like corn.

The average summer temperature in many places is 80°F or more. The winters are cool and moist, with many months having an average temperature of 50°F. The range between the average temperature for the warmest and coolest months is often around 25°F. The growing season is determined by the time between the last frost of the spring and the first frost of the fall.

Just offshore on both continents, a warm ocean current is found. In North America, the current is known as the "Gulf Stream." In South America, the current is known as the "Brazil Current." These are both warm ocean currents coming from the direction of the equator.

The following monthly temperature and rainfall totals are typical of the humid subtropical climate in **North America**.

Month	J	F	M	A	M	J	J	A	S	O	N	D	Year
Temperature (°F)	51	54	60	66	73	79	80	80	77	68	58	52	67
Rainfall (inches)	4.7	5.2	6.4	4.9	4.4	5.4	7.0	7.1	5.3	3.5	3.7	4.9	62.5

The following monthly temperature and rainfall totals are typical of the humid subtropical climate in **South America**.

Month	J	F	M	A	M	J	J	A	S	O	N	D	Year
Temperature (°F)	77	76	70	62	56	49	51	52	57	62	69	75	63
Rainfall (inches)	3.6	3.3	5.0	3.0	2.0	1.5	1.0	1.5	1.5	3.5	3.5	5.0	34.4

Using the above monthly temperature and rainfall charts, complete the following.

1. The three warmest months in the North America chart are   a) May, June, July
    b) August, September, October   c) June, July, August.

2. The months with the most rainfall in the North America chart are   a) July and August,
    b) April and May   c) August and September.

3. The three warmest months in the South America chart are   a) March, April, May
    b) December, January, February   c) October, November, December.

4. The months with the most rainfall in the South America chart are   a) December and March
    b) April and May   c) August and September.

Name: _____ Date: _____

Unit 5: Climate in the Western Hemisphere

D. Humid Subtropical Climate (cont.)

5. The reason why the highest monthly temperatures and rainfall are different between the two locations is because the summer months in the humid subtropical climate of North America are   a) April, May, and June   b) May, June, and July   c) June, July, and August, but the summer months in the humid subtropical climate of South America are

    a) December, January, February   b) October, November, December

    c) February, March, April.

Using **Map 12b**, **12c**, and **an atlas**, complete the following. The number 1s indicate the regions with a humid subtropical climate.

6. For North America, place a "+" on the blank if any part of these states has a humid subtropical climate.

    _____ Alabama   _____ California   _____ Georgia   _____ Wisconsin

    _____ North Carolina _____ South Carolina _____ Virginia

7. For South America, place a "+" on the blank if any part of these countries has a humid subtropical climate.

    _____ Argentina   _____ Chile   _____ Ecuador   _____ Suriname

    _____ Uruguay   _____ Venezuela

DIAGRAM 2

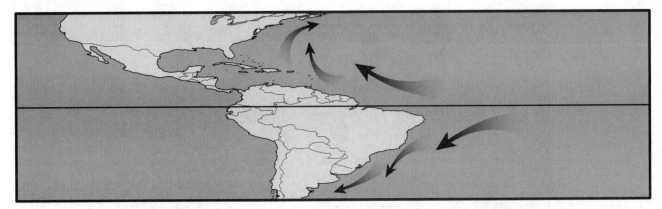

The ocean currents offshore near the humid subtropical climates are shown on **Maps 12b** and **12c** with the lines and arrows as shown in **Diagram 2**.

8. The ocean current offshore along the eastern coast of South America is the

    a) Gulf Stream   b) California   c) Brazil   Current.

9. The ocean current offshore in #8 is a   a) warm   b) cold   ocean current.

10. The ocean current offshore along the east coast of North America is the

    a) Gulf Stream   b) California   c) Brazil   Current.

11. The ocean current offshore in #10 is a   a) warm   b) cold   ocean current.

Name: _____ Date: _____

Unit 5: Climate in the Western Hemisphere

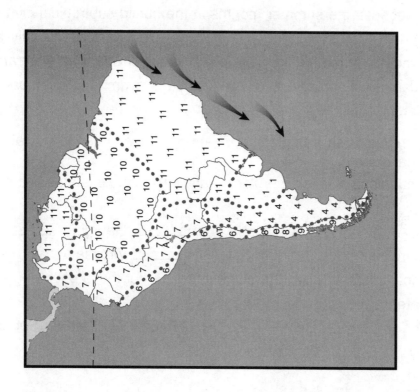

MAP 12c

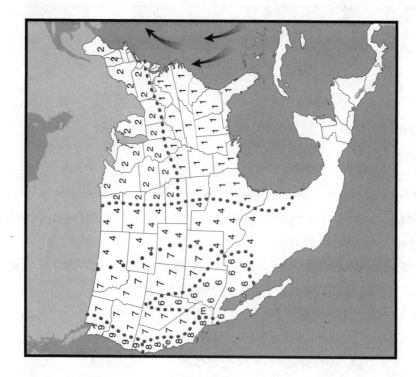

MAP 12b

Unit 5: Climate in the Western Hemisphere

E. Steppe or Semiarid Climate

Where the **steppe** or **semiarid climate** is found, the rainfall is 10 to 20 inches per year. In the United States, most of the region between 100 degrees west longitude and the eastern slope of the Rocky Mountains is steppe. In South America, most of the steppe climate is on the eastern side of the Andes in Argentina. In steppe regions where the rainfall is nearly 20 inches per year, the vegetation is tall grass. However, in steppe areas where the rainfall is ten inches or less, short, sparse grasses become common. Since the steppe regions are located inland away from large bodies of water, the summer days can become very warm, with temperatures reaching 100 degrees. Sheep and cattle are raised on the grasses in this climate.

In both North and South America, the steppe region is on the eastern side of a high mountain range. When the westerlies rise over the western side of the mountains, the airmass cools, and rain or snow falls. As soon as the air starts down the eastern side of the mountains, it warms and picks up moisture, making the region drier.

The vegetation is usually short grasses. The area on the east side of the Rocky Mountains and Andes Mountains is called the "rain shadow."

DIAGRAM 3

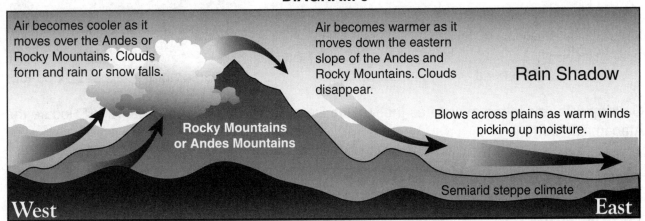

Air becomes cooler as it moves over the Andes or Rocky Mountains. Clouds form and rain or snow falls.

Air becomes warmer as it moves down the eastern slope of the Andes and Rocky Mountains. Clouds disappear.

Rain Shadow

Blows across plains as warm winds picking up moisture.

Rocky Mountains or Andes Mountains

Semiarid steppe climate

West

East

Using **Maps 12a**, **12b**, **12c**, and **an atlas**, complete the following activity.

1. The number 4s indicate the regions with a semiarid or steppe climate. Draw a line around the 4s to show this.
2. The semiarid, or steppe, climate is found in   a) Argentina, Uruguay, Ecuador
   b) Argentina, the United States, Canada, and British Guiana
   c) Argentina, the United States, Canada, and Mexico.
3. True or False. A large part of the Great Plains in the United States has a semiarid climate.
4. True or False. The semiarid climate is found on the Patagonia Plateau in Argentina.
5. True or False. The semiarid climate is most commonly found near the coast.
6. True or False. The semiarid climate is most commonly found on the western side of the Andes and the Rocky Mountains.
7. True or False. Large areas of semiarid climate are found inland away from the coast.
8. True or False. In both North and South America, sheep and cattle are raised in regions with a semiarid climate.

51

Name: _____ Date: _____

Unit 5: Climate in the Western Hemisphere

F. Desert Climate

The **desert climate** is found in the southwestern region of North America, in the Patagonia Plateau in Argentina, and along the Pacific Coast of Chile, Ecuador, and Peru in South America. The desert climate has very hot summers. Rainfall is less than ten inches per year, and little, if any, vegetation is found.

Using **Map 12b**, **12c**, and **an atlas**, complete the following. The number 6s indicate the desert climate.

1. Place a "+" on the blank if a desert climate is found in any part of the following states.

   _____ Arizona   _____ Iowa

   _____ Maine   _____ Missouri   _____ New Mexico

   _____ Nevada

2. Place a "+" on the blank if a desert climate is found in any part of these South American countries.

   _____ Argentina   _____ Chile   _____ Colombia   _____ Ecuador

On **Map 12b**, the letter "m" locates the Mojave Desert. On **Map 12c**, the letters "AT" locate the Atacama Desert.

3. The Atacama Desert is a very dry desert with no rain falling for years. It is located in

   a) Brazil   b) Texas   c) Chile   d) Venezuela

4. The Mojave Desert is located on the continent of   a) North America   b) South America.

5. The Mojave Desert is located in   a) Chile   b) Uruguay   c) California.

G. Highland Climate

A **highland climate** is found in both North and South America. The highland climate is marked by much cooler temperatures than surrounding lower elevations. For each one thousand feet of elevation, the temperature becomes three degrees cooler. As the temperature becomes cooler, the vegetation changes.

In South America, many people live on the high plateaus in the mountains. People living on the plateaus are thousands of feet above those who live along the Pacific coast. The plateaus are surrounded by high mountains.

Along the Pacific coast in South America, the climate is much warmer than is found on the highland plateaus. In some regions, the coastal area is desert, while along other regions the coastal areas are wetter and more humid.

Name: _____ Date: _____

Unit 5: Climate in the Western Hemisphere

G. Highland Climate (cont.)

In **Diagram 4** below, imagine the letter "A" is a city along the coast of Peru and the letter "B" is a city on the high plateau in Peru. Imagine the high, snow-capped mountain is Mt. Coattail, which is over 29,000 feet high.

A. Elevation 100 feet or near sea level. Average temperature for coldest month 77°F. Average temperature for warmest month 82°F. Average yearly rainfall 34 inches.
B. Elevation 9,000 feet. Average temperature for coldest month 54°F. Average temperature for warmest month 65°F. Average yearly rainfall 20 inches.

DIAGRAM 4

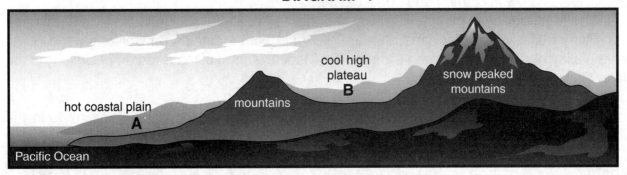

Using **Diagram 4** above, answer the following questions.
1. In traveling from City A to City B, the average monthly temperature becomes much
    a) warmer   b) cooler.
2. The average temperature decreases by three degrees for each 1,000 feet of elevation. If the average temperature in City A is 82°F, and the elevation is 0, you would expect the average temperature in City B at an elevation of 9,000 feet to be
    a) 72°F   b) 40°F   c) 82°F   d) 55°F.

In South America, many people live on the plateaus in the high Andes Mountains. One famous plateau is the Altiplano. People have lived there for hundreds of years. The main center for the great Inca civilization was located there. Two large lakes located on the Altiplano are Lake Poopó and Lake Titicaca.

Using **Map 12b**, **12c**, and **an atlas**, complete the following activity. You will need a purple colored pen or pencil for this activity.
1. Number 7s indicate the region with a highland climate. Color the regions with 7s purple.
2. The letters "AP" indicate the location of the high Altiplano. It is located in the countries of
    a) Venezuela and Peru   b) Peru and Bolivia   c) Argentina and Paraguay.
3. Lake Poopó and Lake Titicaca are identified with a gray shaded lake symbol near the Altiplano. These lakes are located in   a) Venezuela and Brazil
    b) Colombia and Peru   c) Peru and Bolivia   d) Argentina and Paraguay.
4. Place a "+" on the blank if a highland climate is found in any part of these states.
    _____ Colorado   _____ Idaho   _____ Illinois   _____ Oregon
    _____ Washington   _____ Wyoming

Name: _____ Date: _____

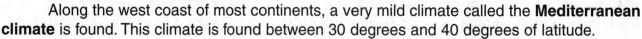

Unit 5: Climate in the Western Hemisphere

H. Mediterranean Climate

Along the west coast of most continents, a very mild climate called the **Mediterranean climate** is found. This climate is found between 30 degrees and 40 degrees of latitude.

The Mediterranean climate found near the coastal areas in California and Chile has very hot, dry summers, with the rainfall coming in the winter. In summer, skies are usually very clear with few clouds, if any. The daytime temperatures can become very high. However, nights can be rather cool, as the clear skies allow the earth to lose its temperature rapidly. The winter temperatures are very mild and sunny. However, most of the rainfall does come in the winter months. The vegetation in the Mediterranean climate consists of shrubs, bushes, and small trees. Grasses grow in clumps and small patches.

The Mediterranean climate is a very important agricultural region in both California and Chile. Many fruits, vegetables, and nuts are grown in both locations. In California, the Great Central Valley has a Mediterranean climate. The Mediterranean climate region of Chile is in the Vale of Chile.

The following monthly temperature and rainfall totals are for a Mediterranean climate in the **Northern Hemisphere**.

Month	J	F	M	A	M	J	J	A	S	O	N	D	Year
Temperature (°F)	49	51	53	54	56	57	57	58	60	59	56	51	55
Rainfall (inches)	4.8	3.6	3.1	1.0	0.7	0.1	0.0	0.0	0.3	1.0	2.4	4.6	22.2

The following monthly temperature and rainfall totals are for a Mediterranean climate in the **Southern Hemisphere**.

Month	J	F	M	A	M	J	J	A	S	O	N	D	Year
Temperature (°F)	57	58	60	59	56	51	49	51	53	54	56	57	55
Rainfall (inches)	0.0	0.0	0.3	1.0	2.4	4.6	4.8	3.6	3.1	1.0	0.7	0.1	21.6

Using **Maps 12b**, **12c**, and **an atlas**, complete the following activity.

The letter "e" locates the Vale of Chile and the Great Central Valley.

1. Write the letters "VC" on the Vale of Chile.

2. The Vale of Chile is located in _____ on the continent of _____.

3. Write the letters "GVC" on the Great Central Valley.

4. The Great Central Valley is located in _____ on the continent of _____.

Unit 5: Climate in the Western Hemisphere

H. Mediterranean Climate (cont.)

5. In the Mediterranean climate, the rainfall comes in the   a) winter   b) summer.
6. In the Mediterranean climate, the   a) summers   b) winters   are dry.
7. In the Vale of Chile, the rainfall comes in   a) summer   b) fall   c) winter   d) spring.
8. In the Great Central Valley, the rainfall comes in   a) summer   b) fall   c) winter

    d) spring.
9. Winter in the Great Central Valley is   a) December, January, February

     b) June, July, August   because the Great Central Valley is located in the

     a) Southern   b) Northern   Hemisphere.
10. Winter in the Vale of Chile is in a) December, January, February   b) June, July, August

     because the Vale of Chile is located in the a) Southern b) Northern Hemisphere.

Using **Maps 12b**, **12c**, and **an atlas**, complete the following activity. You will need a green colored pen or pencil for this activity. The regions with the number 8s have a Mediterranean climate.

1. Color the regions with the number 8s inside the dotted lines green.
2. The Mediterranean climate is found on the (east, west) coast of both North America and South America.
3. The Mediterranean climate in South America is found between _____ degrees south and _____ degrees south latitude.
4. The Mediterranean climate in North America is found between _____ degrees north and _____ degrees north latitude.
5. The Mediterranean climate in North America is found in   a) Texas   b) California

     c) Illinois   d) Washington.
6. The Mediterranean climate in South America is found in   a) Brazil   b) Argentina

     c) Chile   d) Colombia.
7. The ocean current flowing along the coast of Chile is the   a) Brazil   b) Peru

     c) California d) Gulf Stream   Current.
8. The ocean current flowing along the coast of California is the a) Brazil   b) Peru

     c) California d) Gulf Stream   Current.
9. Both ocean currents are coming from the direction of the Poles, so both are

     a) warm currents   b) cold currents.
10. The offshore winds blow   a) toward the land   b) toward the ocean.
11. The winds are the   a) northeast trades   b) southeast trades

     c) polar easterlies   d) westerlies.

Name: _____ Date: _____

Unit 5: Climate in the Western Hemisphere

I. West Coast Marine Climate

The **west coast marine climate** is found on the western coasts of both North America and South America. In North America, this climate extends along the Pacific Ocean from 40 degrees north latitude north along the coast of Alaska. This is the wettest climate in the continental United States, with many locations receiving more than 60 inches of rainfall each year. Summers are cool, and winters are mild but wet. Many days are cloudy with rain.

In South America, the west coast marine climate begins at 35 degrees south latitude and extends south to the tip of Cape Horn. Like the climate in North America, summers are cool, and winters are mild. Rainfall can be as much as 50 to 60 inches per year. The rainfall comes throughout the year.

The westerly winds blow onshore across the California Current in North America and the Peru Current in South America. The westerly winds and warm ocean current make the winters milder than one would expect at latitudes where the west coast marine climate is found.

In both North America and South America near the Pacific coast are high mountains. The westerly winds blow across the ocean currents, which are warmer than the surrounding water, and begin to rise over the mountains. When the airmass begins to move over the mountains, it is cooled. As it cools, clouds form, and rain begins to fall. When the airmass rises higher, the rain changes to snow. Rain and snow that result from an airmass rising over high mountains is called **orographic precipitation**.

Because of the mild, wet climate, there are great forests in the west coast marine climate. In both North and South America, the lumber industry is very important in this climate.

DIAGRAM 5

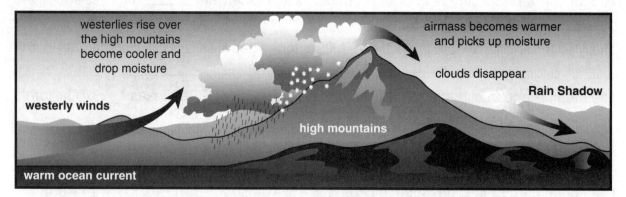

1. **Using Diagram 5 above,** select the answer below that best explains **orographic precipitation**. (Read very carefully.)

   a) The westerly winds blow onshore across the warm ocean current, begin to rise over the mountains, and become warmer. Finally, precipitation will fall as rain or snow.

   b) The westerly winds blow onshore across the warm ocean current, begin to rise over the mountains and become cooler, clouds form, and drop precipitation as rain or snow.

   c) The westerly winds blow offshore across the warm ocean current, begin to rise over the mountains, and become cooler and drop precipitation as rain or snow.

Name: _____ Date: _____

Unit 5: Climate in the Western Hemisphere

I. West Coast Marine Climate

Using **Maps 12a**, **12b**, **12c**, and **an atlas**, complete the following. You will need a blue colored pen or pencil for this activity.

1. The west coast marine climate is indicated by the number 9s. Color the region with the number 9s blue.

2. In North America, the west coast marine climate is found along the Pacific coast north of
    a) 30 degrees north latitude   b) 20 degrees north latitude   c) 40 degrees north latitude.

3. In South America, the west coast marine climate is found along the Pacific coast south of
    a) 30 degrees south latitude   b) 35 degrees south latitude   c) 10 degrees south latitude.

4. In the United States, the west coast marine climate is found in
    a) Washington and Oregon   b) Idaho and Oregon   c) Washington and Utah.

5. In South America, the west coast marine climate is found in   a) Brazil and Chile
    b) Chile and Argentina   c) Venezuela and Colombia.

The following monthly temperature and rainfall totals are for a west coast marine climate.

Month	J	F	M	A	M	J	J	A	S	O	N	D	Year
Temperature (°F)	39	42	46	51	57	61	67	66	61	54	46	41	53
Rainfall (inches)	7.0	6.0	5.0	3.0	2.5	1.5	0.5	0.5	2.0	3.0	6.5	7.0	44.5

Complete the following with a bar graph to show the rainfall and a line graph to show the temperature for the above location.

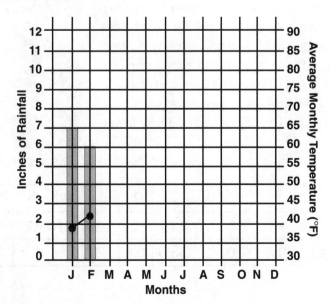

Name: _____ Date: _____

Unit 5: Climate in the Western Hemisphere

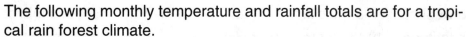

J. Tropical Rain Forest Climate

In the Amazon River region of South America, the **tropical rain forest climate** is found. Tall trees form a canopy over the ground below. The average monthly temperature is in the high 70s to low 80s. The daytime temperatures are usually in the 90s, with nighttime temperatures in the 70s. In the afternoon, temperatures rise, and as the warm air near the earth's surface begins to rise, thunderstorms develop, bringing heavy downpours of rain.

The following monthly temperature and rainfall totals are for a tropical rain forest climate.

Month	J	F	M	A	M	J	J	A	S	O	N	D	Year
Temperature (°F)	78	78	76	77	76	74	74	76	76	77	78	78	76.5
Rainfall (inches)	10	10	12	7	10	7	7	5	9	7	8	11	103

K. Tropical Savanna (Wet and Dry)

In South America on the northern and southern boundaries of the tropical rain forest climate, the **tropical savanna** is found. The tropical savanna has a definite wet season and a definite dry season. The wet season comes during the summer; the dry season comes during the winter. Although the seasons are spoken of as summer and winter, the tropical savanna is located so close to the equator that temperatures remain warm throughout the year. During the wet season, rain falls most days and is usually heavy. Lakes are full, and rivers are overflowing. During the dry season, the rains stop, and the lakes and rivers become dry. The vegetation in the savanna climate is mostly tall grasses. However, near the boundary with the tropical rain forest, there are many trees.

The following monthly temperature and rainfall totals are for a savanna climate.

Month	J	F	M	A	M	J	J	A	S	O	N	D	Year
Temperature (°F)	76	76	77	78	79	80	81	81	82	82	81	80	79.5
Rainfall (inches)	1.0	0.5	1.0	1.1	3.9	3.8	3.0	3.5	4.0	6.0	3.0	1.0	31.8

58

Name: _____ Date: _____

Unit 5: Climate in the Western Hemisphere

J. Tropical Rain Forest Climate/K. Tropical Savanna Climate (Wet and Dry)

Using **Map 12c** and **an atlas**, complete the following. You will need a red and a pink colored pen or pencil for this activity.

1. Draw a line to connect the dashes.
2. Label the line "Equator, 0°."
3. The tropical rain forest climate is indicated by the number 10s. Color this region red.
4. The tropical savanna climate is indicated by the number 11s. Color this region pink.
5. North and south of the tropical rain forest the   a) Mediterranean   b) desert
    c) tropical savanna   d) west coast marine   climate is found.
6. The tropical rain forest is found in   a) Brazil   b) Argentina   c) Texas   d) Chile.
7. The tropical savanna is found in   a) Brazil   b) Ecuador   c) Texas   d) Chile.

Use the monthly temperature and rainfall information for the tropical rain forest and tropical savanna on the previous pages to complete the following.

8. The   a) tropical savanna   b) tropical rain forest   is found nearest the equator.
9. The tropical savanna is found   a) east and west   b) north and south   of the tropical rain forest.
10. The   a) tropical savanna   b) tropical rain forest   has a definite wet season and a definite dry season.
11. The large river that is found in the tropical rain forest is the   a) Mississippi b) Paraná   c) Amazon   d) Orinoco.
12. The   a) tropical rain forest   b) tropical savanna   is a climate where large forests of trees are found.

Name: _____ Date: _____

Unit 5: Climate in the Western Hemisphere

L. Identifying Climates

To complete the following chart, you will need maps that show the climates of North America and South America.

1. Place each climate term in the correct column(s) where that climate is found.

desert	highland	humid continental
humid subtropical	Mediterranean	steppe
subarctic	tropical rain forest	tropical savanna
tundra	west coast marine	

North America	South America

2. A climate type that is found in South America but not in North America is

   a) Mediterranean   b) tropical rain forest   c) tundra   d) desert.

3. Climate types that are found only in North America are

   a) west coast marine and Mediterranean   b) tropical rain forest and tropical savanna

   c) tundra and subarctic   d) steppe and desert.

4. Climate types that are found in both North America and South America are

   a) tropical rain forest and tropical savanna   b) tundra and subarctic

   c) steppe and desert.

Name: _____ Date: _____

Unit 5: Climate in the Western Hemisphere

M. Pretest Practice

Complete the following using the terms below. Some terms may be used more than once.

California	**eastern**	**highland**	**humid subtropical**
Mediterranean	**west coast marine**	**orographic**	**Peru**
three	**tropical rain forest**	**tundra**	

1. The type of climate found on the Altiplano of Bolivia and Peru is _____.

2. The type of climate found in the Yukon Territory and Northwest Territories is the

   _____.

3. The climate type found on the southeast coast of North America and in South America is

   _____.

4. The climate type found on the west coast of North America and South America between

   30° and 40° latitude is _____.

5. The climate type found in Brazil near the Amazon River is _____.

6. The climate type found on the west coast of North America and in South America at lati-

   tudes above 40 is _____.

7. The ocean current near Chile is the _____ Current.

8. The ocean current near Washington and Oregon is the _____

   Current.

9. The tropical savanna climate is found on the northern and southern borders of the

   _____ climate.

10. Rainfall or snow that falls when an airmass moves over high mountains is

    _____ precipitation.

11. When climbing higher over a mountain, for each 1,000 feet of elevation, the temperature

    will become _____ degrees cooler.

12. In both North America and South America, desert and semiarid climates are often found

    on the _____ side of the Andes and Rocky Mountains.

Name: _____ Date: _____

Unit 5: Climate in the Western Hemisphere

M. Test

Circle the letter of the correct answer.

1. The type of climate found on the Altiplano of Bolivia and Peru is   a) highland
    b) west coast marine   c) tundra   d) humid subtropical.

2. In the Western Hemisphere, the type of climate found in the Yukon Territory and Northwest
    Territories is   a) highland   b) west coast marine   c) tundra   d) humid subtropical.

3   The climate type found on the southeast coast of North America and in South America is
    a) Mediterranean b) humid continental c) tundra d) humid subtropical.

4. The climate type found on the west coast of North America and in South America between
    30° and 40° latitude is   a) Mediterranean   b) humid continental   c) tundra
    d) humid subtropical.

5. The climate type found near the Amazon River in South America is   a) tropical savanna
    b) tropical rain forest   c) subarctic d) mid-latitude marine.

6. The climate type found on the west coast of North America and in South America between
    latitudes 40° and 60° is   a) Mediterranean   b) west coast marine   c) steppe   d) desert.

7. The ocean current off the coast of Chile is the   a) Peru b) Brazil c) Gulf Stream
    d) California   Current.

8. The ocean current off the coast of Washington and Oregon is the   a) Peru
    b) Brazil   c) Gulf Stream   d) California   Current.

9. The tropical savanna climate is found on the northern and southern borders of the
    a) Mediterranean   b) humid subtropical c) steppe   d) tropical rain forest   climate.

10. Rain or snow that falls when an air mass moves over high mountains is called
    a) orographic   b) rain shadow   c) periodic   precipitation.

11. When climbing higher over a mountain, for each 1,000 feet of elevation, the temperature
    will become approximately   a) 3°   b) 8°   c) 1°   d) 10°   cooler.

12. In both North America and South America, desert and semiarid climates are often found
    on the   a) western   b) northern   c) eastern   d) southern   side of
    the Andes and the Rocky Mountains.

Name: _____ Date: _____

Unit 6: Major Cities of the Western Hemisphere

Using **Map 13** and **an atlas**, complete the following. Each number and dot on the maps locate a city. These numbers correspond with the cities listed in the chart below. Write the name of the city on the blank by the corresponding dot on the maps. Place a "+" in the chart to show the location of the city as North America or South America. Place a "+" in the chart to show that the city has an inland or coastal location.

	CONTINENT		LOCATION	
City	North America	South America	Coastal	Inland
1. Mexico City				
2. Seattle				
3. Boston				
4. Caracas				
5. Bogotá				
6. Buenos Aires				
7. Detroit				
8. Buffalo				
9. Philadelphia				
10. Washington, D.C.				
11. Cleveland				
12. Baltimore				
13. São Paulo				
14. Guayaquil				
15. Belo Horizonte				
16. Miami				
17. New Orleans				
18. Houston				
19. Chicago				
20. St. Louis				
21. San Francisco				
22. Milwaukee				
23. Minneapolis				
24. Denver				
25. Dallas				
26. Los Angeles				
27. Kansas City				
28. Cincinnati				
29. Belém				
30. Recife				
31. Medellín				
32. Montreal				
33. Toronto				
34. New York City				

63

Unit 6: Major Cities of the Western Hemisphere

MAP 13

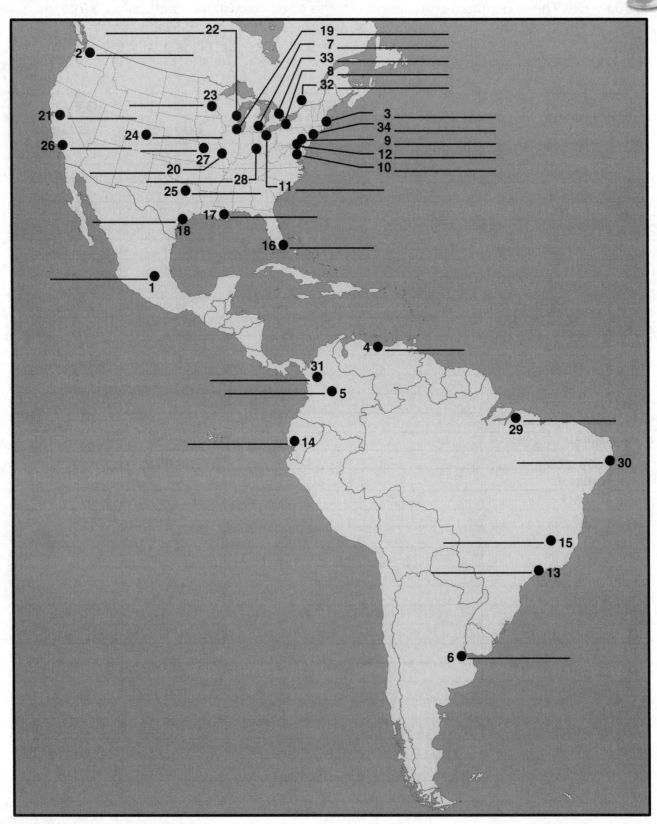

Name: _____ Date: _____

Unit 6: Major Cities of the Western Hemisphere

Use **Map 13** and/or **an atlas** to complete the following exercise.

1. If the city has a coastal or near-coastal location; place the letter "C" on the blank. If the city has an inland location; place the letter "I" on the blank. (Populations are listed in millions (m) in parentheses.)

__ a. Belo Horizonte (1.5m)   __ b. Bogotá (4m)   __ c. Buenos Aires (3m)
__ d. Caracas (1.8m)   __ e. Chicago (2.8m)   __ f. Guadalajara (1.7m)
__ g. Havana (2.1m)   __ h. Houston (1.6m)   __ i. Los Angeles (3.5m)
__ j. Mexico City (8.2m)   __ k. New York City (7.3m)   __ l. Philadelphia (1.6m)
__ m. Rio de Janeiro (5.5m)   __ n. Salvador (2.1m)   __ o. Santo Domingo (2.4m)
__ p. São Paulo (9.4m)

2. Most large cities in South America are located

    a) near the coast   b) inland   c) near the coast and inland.

3. Most large cities in North America are located

    a) near the coast   b) inland   c) near the coast and inland.

4. Complete the bar graph below for the ten most populated cities in the Western Hemisphere using the population numbers above.

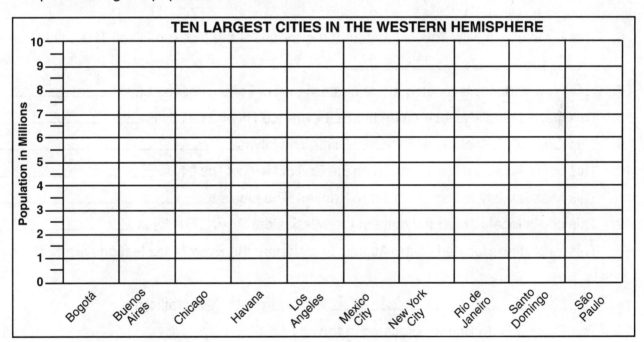

Use the bar graph above to answer the following questions.

5. The largest city in North America is a) New York City b) Mexico City c) Philadelphia.

6. The largest city in South America is a) Buenos Aires b) Rio de Janeiro c) São Paulo.

7. The largest city in South America is located in a) Brazil b) Argentina c) Venezuela.

Name: _____ Date: _____

Unit 6: Major Cities of the Western Hemisphere—Pretest Practice

Complete the following with the names of the cities below. Use an atlas or other reference sources.

Baltimore	**Belém**	**Belo Horizonte**	**Bogotá**	**Boston**
Buenos Aires	**Buffalo**	**Caracas**	**Chicago**	**Cincinnati**
Cleveland	**Dallas**	**Denver**	**Detroit**	**Guayaquil**
Houston	**Kansas**	**Los Angeles**	**Medellín**	**Mexico City**
Miami	**Milwaukee**	**Minneapolis**	**Montreal**	**New Orleans**
New York City	**Philadelphia**	**Recife**	**San Francisco**	**São Paulo**
Seattle	**St. Louis**	**Toronto**	**Washington, D.C.**	

1. This city is a seaport city in Peru. The city is _____.

2. The largest city in Illinois, located on Lake Michigan, is _____.

3. A national capital located on the Potomac River is _____.

4. A large Canadian city located on Lake Ontario is _____.

5. This city in Minnesota is located near where the Mississippi River begins. The city is

   _____.

6. This large city in New York is located at the mouth of the Hudson River. The city is

   _____.

7. This New York city is located at the eastern end of Lake Erie. The city is _____.

8. This Canadian city is located on the St. Lawrence River. The city is _____.

9. This Ohio city is located on the Ohio River. The city is _____.

10. This city in Texas is a large seaport on the Gulf of Mexico. The city is _____.

11. This Missouri city is on the border of Kansas. The city is _____.

12. This city is located near the delta of the Mississippi River. The city is _____.

13. This large city is located on the Atlantic Ocean near the Everglades National Park. The city

    is _____.

14. _____ is the largest city in California.

15. The Golden Gate Bridge is located in the city of _____.

16. This Wisconsin city is located on Lake Michigan. The name of the city is _____.

17. The city of _____ is known as the "City of Brotherly Love."

18. This city is located on a high plateau. Mt. Popocatépetl can be seen in the distance. The

    city is _____.

66

Name: _____ Date: _____

Unit 6: Major Cities of the Western Hemisphere—Pretest Practice (cont.)

19. This city in Colombia is located in the mountains. It is about five degrees north of the equator. The city is _____.

20. This city is located approximately 11 degrees north of the equator in the mountains of Venezuela. The city is _____.

21. This city is on the Mississippi River. It is near the point where the Missouri River flows into the Mississippi. It is known as the "Gateway to the West." The city is _____.

22. This large automobile manufacturing city is located on the border with Canada. It is located on Lake St. Clair. The city is _____.

23. This city is located in the Rocky Mountains and is located at approximately 39 degrees north latitude. This city is _____.

24. This city is located on the Atlantic Coast near Cape Cod. Near here at Breed's Hill and Bunker Hill, the colonists resisted the British. The city is _____.

25. This city is located at the southern end of Puget Sound. The city is _____.

26. This city is located at the mouth of the Amazon River. The city is _____.

27. Located on the Atlantic Ocean at approximately 23° south latitude, this city is famous for its beautiful beaches. The city is _____.

28. _____ is located at the mouth of the Parana River and on the Rio de la Plata.

29. _____ is located farther east than any other Brazilian city. It is located on the Atlantic Ocean. The city is _____.

30. This Ohio city is located on Lake Erie. North across Lake Erie is the Ontario Peninsula of Canada. The city is _____.

31. This city is located in Maryland on the Chesapeake Bay. The city is _____.

32. This city is located approximately $23\frac{1}{2}°$ degrees south on the Tropic of Capricorn. The city is _____.

33. _____ is a port city located on the Pacific Ocean in Ecuador.

34. Located at approximately 20 degrees south latitude, this city in Brazil is noted for its iron and steel manufacturing. The city is _____.

Name: _____ Date: _____

Unit 6: Major Cities of the Western Hemisphere—Test

Circle the letter of the correct answer.

1. This city is a seaport city in Peru. The city is a) Lima b) Guayaquil c) Belém
    d) Buenos Aires.
2. The largest city in Illinois and located on Lake Michigan is   a) Chicago b) Los Angeles
    c) St. Louis   d) Toronto.
3. A national capital located on the Potomac River is   a) San Francisco
    b) Washington, D.C.   c) New Orleans   d) Los Angeles.
4. A large Canadian city located on Lake Ontario is   a) Montreal   b) Detroit
    c) Toronto   d) Milwaukee.
5. This large city in New York is located at the mouth of the Hudson River. The city is
    a) Philadelphia   b) Boston   c) Chicago   d) New York City.
6. This Canadian city is located on the St. Lawrence River. The city is   a) Montreal
    b) Toronto   c) Cleveland   d) Buffalo.
7. This city in Texas is a large seaport on the Gulf of Mexico. The city is a) Dallas
    b) Houston   c) New Orleans   d) Miami.
8. This city is located on the Mississippi delta near where the river flows into the Gulf of
    Mexico. The city is   a) Houston   b) St. Louis c) Kansas City   d) New Orleans.
9. This large city is located on the Atlantic Ocean near the Everglades National Park. The city
    is   a) Miami   b) New Orleans   c) Boston   d) New York City.
10. This city is the largest city in California. The city is a) San Francisco b) Seattle
    c) Los Angeles   d) Denver.
11. The Golden Gate Bridge is located in the city of   a) San Francisco   b) Seattle
    c) Los Angeles   d) Denver.
12. This United States city is known as the "City of Brotherly Love." The city is   a) Boston
    b) New York City   c) Philadelphia   d) Washington, D.C.
13. This city is located on a high plateau. Mt. Popocatépetl can be seen in the distance. The
    city is   a) Mexico City   b) Caracas   c) Medellín   d) Belém.
14. This city on the Mississippi River, known as the "Gateway to the West," is located near the
    point where the Missouri River flows into the Mississippi River. The city is
    a) Kansas City   b) St. Louis c) New Orleans   d) Chicago.
15. This city is located on the Atlantic Coast near Cape Cod. Near here at Breed's Hill and
    Bunker Hill, the colonists resisted the British. The city is   a) Boston
    b) Washington, D.C.   c) Philadelphia   d) New York City.
16. This city is located at the mouth of the Amazon River. The city is   a) São Paulo
    b) Belo Horizonte   c) Brasília   d) Belém.
17. This city located in Brazil on the Atlantic Ocean is famous for its beautiful beaches. It is
    located at approximately 23 degrees south latitude. The city is   a) Belém
    b) Caracas   c) Rio de Janeiro   d) Recife.
18. This South American city is located at the mouth of the Parana River and is located on the
    Rio de la Plata. The city is   a) São Paulo b) Rio de Janeiro c) Belem d) Buenos Aires.
19. This city is located in Maryland on the Chesapeake Bay. The city is   a) Boston
    b) Miami   c) Baltimore   d) Detroit.
20. Located at approximately 20° south latitude, this city in Brazil is noted for its iron and steel
    manufacturing. The city is   a) Belo Horizonte   b) Caracas   c) Belém   d) Brasília.

Name: _____ Date: _____

Unit 7: Agriculture in the Western Hemisphere

A. Tropical Crops

The tropics are a region of year-round high temperatures. The tropical rain forest, tropical savanna, and highland climates are found there. In the Central American and West Indies island nations, the crops are often raised on large plantations.

Bananas, oranges, pineapples, lemons, cacao beans, sugar cane, rice, tobacco, and coffee are major crops that are grown in the tropics. The tropics are found between $23\frac{1}{2}°$ north latitude and $23\frac{1}{2}°$ south latitude.

Using **Map 14** and **an atlas**, complete the following. You will need a green colored pen or pencil for this activity.

1. A dotted line marks $23\frac{1}{2}°$ north latitude and $23\frac{1}{2}°$ south latitude. Shade the area between the dotted lines green and write "tropics" in the shaded area.

2. Place a "+" on the blank if a majority of the country is located in the tropics.

   _____ Argentina   _____ Brazil   _____ Canada

   _____ Chile   _____ Costa Rica   _____ Cuba

   _____ Dominican Republic _____ El Salvador   _____ Guatemala

   _____ Haiti   _____ Honduras   _____ Jamaica

   _____ Panama   _____ Puerto Rico   _____ Trinidad

   _____ United States   _____ Uruguay

3. Place a "+" on the blank if the country is an island nation located in the tropics.

   _____ Belize   _____ Costa Rica   _____ Cuba

   _____ Dominican Republic _____ El Salvador   _____ Honduras

   _____ Jamaica   _____ Panama   _____ Puerto Rico

   _____ Trinidad

4. Place a "+" on the blank to identify major crops grown on large plantations in the tropics.

   _____ bananas   _____ coffee   _____ corn   _____ oats

   _____ oranges   _____ rice   _____ sugar cane   _____ tobacco

   _____ wheat

5. Place a "+" on the blank if tropical crops are grown there.

   _____ Bolivia   _____ Canada   _____ Chile   _____ Colombia

   _____ Costa Rica _____ Cuba   _____ Haiti   _____ Honduras

   _____ Jamaica   _____ Nicaragua

Name: _____ Date: _____

Unit 7: Agriculture in the Western Hemisphere

A. Tropical Crops (cont.)

MAP 14

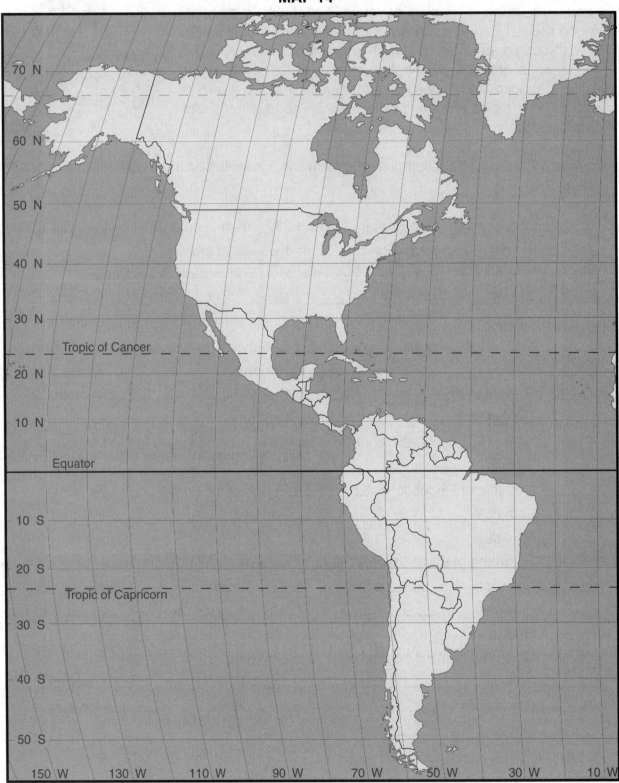

Name: _____ Date: _____

Unit 7: Agriculture in the Western Hemisphere

A. Tropical Crops—Pretest Practice

Complete the following using the terms below. Some terms may be used more than once.

$23\frac{1}{2}°$	bananas	Caribbean	Central America
cacao	coffee	Cuba	highland
Honduras	rice	sugar cane	soybeans
Trinidad	rain forest	savanna	Tropic of Cancer
Tropic of Capricorn	tropics	West Indies	

1. The tropics are located between _____ north latitude and _____ south latitude.

2. A major fruit crop raised on large plantations in the tropics is _____.

3. A term used to include all of the Caribbean island nations is the _____.

4. A term used to include the nations on the isthmus between North and South America that are not island nations is _____.

5. A major crop grown in the highlands on large plantations in Colombia, Mexico, and Brazil and shipped to the United States where it is a popular drink is _____.

6. The line of latitude that is $23\frac{1}{2}°$ north is known as the Tropic of _____.

7. The line of latitude that is $23\frac{1}{2}°$ south is known as the Tropic of _____.

8. A major food crop grain that is grown on large plantations in the tropics is _____.

9. The _____ bean is grown on trees in the tropics and is used to make chocolate.

10. The island of _____ is a major tobacco producer.

11. The climate in the tropics with high year-round temperatures and rainfall is the tropical _____ climate.

12. A climate in the tropics with high year-round temperatures but having a dry season and a wet season is the tropical _____.

13. A cool climate found in the mountain areas where the coffee plantations are found is the _____ climate.

14. A Central American country with large banana plantations is _____.

15. The West Indies are located in the _____ Sea.

16. The region between $23\frac{1}{2}°$ north latitude and $23\frac{1}{2}°$ south latitude is known as the _____.

17. A major plantation crop raised on the islands in the West Indies and used to sweeten drinks is _____.

Name: _____ Date: _____

Unit 7: Agriculture in the Western Hemisphere

A. Tropical Crops—Test

Circle the letter of the correct answer.

1. The tropics are located between a) $23\frac{1}{2}°$ b) 45°   north latitude and   c) $23\frac{1}{2}°$ d) 45° south latitude.

2. A major fruit crop raised on large plantations in the tropics is   a) corn   b) wheat c) bananas.

3. A term used to include all of the Caribbean island nations is   a) Central America b) South America   c) North America   d) the West Indies.

4. A term used to include the nations that are located on the isthmus between North America and South America is   a) the West Indies b) Central America c) South America.

5. A major crop grown in the highlands on large plantations in Colombia, Mexico, and Brazil and shipped to the United States where it is a popular drink is a) coffee b) yerba maté.

6. The line of latitude that is $23\frac{1}{2}°$ north is known as the Tropic of   a) Cancer   b) Capricorn.

7. The line of latitude that is $23\frac{1}{2}°$ south is known as the Tropic of   a) Cancer b) Capricorn.

8. A major food crop grain that is grown on large plantations in the tropics is
   a) apples   b) coffee   c) rice   d) corn.

9. The bean grown on trees in the tropics and used to make chocolate is   a) coffee b) cacao   c) tea   d) soybeans.

10. A major tobacco producer is the island nation of   a) Dominican Republic   b) Haiti c) Trinidad   d) Galapagos.

11. A climate in the tropics with high year-round temperatures and rainfall is the
    a) west coast marine   b) Mediterranean   c) tropical savanna   d) tropical rain forest.

12. A climate in the tropics with high year-round temperatures but having a dry season and a wet season is the a) highland b) Mediterranean c) tropical savanna d) tropical rain forest.

13. A cool climate found in the mountain areas where the coffee plantations are found is the
    a) highland b) Mediterranean c) tropical savanna d) tropical rain forest.

14. A Central American country with large banana plantations is   a) Haiti   b) Cuba c) Honduras   d) Colombia.

15. The West Indies are located in the a) Coral   b) Mediterranean   c) Caribbean d) Arctic Sea.

16. The region between $23\frac{1}{2}°$ north latitude and $23\frac{1}{2}°$ south latitude is known as the
    a) mid-latitudes   b) tropics   c) polar latitudes.

17. A major plantation crop raised on the islands in the West Indies and used to sweeten drinks is   a) sugar cane   b) tobacco   c) rice   d) cacao.

Name: _____ Date: _____

Unit 7: Agriculture in the Western Hemisphere

B. Mid-Latitude Crops

The mid-latitudes are found between 30° and 60° north latitude and 30° and 60° south latitude. Many crops are grown in the mid-latitudes; these include apples, plums, pears, peaches, oranges, grapefruit, grapes, rice, corn, soybeans, oats, wheat, cotton, and tobacco. Beef cattle, hogs, sheep, and dairy animals are raised in large numbers, primarily for food.

Using **Map 14** and **an atlas**, complete the following.

1. Lines mark 30° and 60° north latitude and 30° and 60° south latitude. Shade the areas between each set of lines and write "mid-latitudes" in the shaded areas.

2. Place a "+" on the blank if a majority of the country is located in the mid-latitudes.

_____ Argentina	_____ Brazil	_____ Canada
_____ Chile	_____ Cuba	_____ Dominican Republic
_____ Haiti	_____ Honduras	_____ Trinidad
_____ United States	_____ Uruguay	_____ Venezuela

3. Place a "+" on the blank if the country grows mid-latitude crops.

_____ Argentina	_____ Canada	_____ Chile	_____ Colombia
_____ Costa Rica	_____ Cuba	_____ Honduras	_____ Jamaica
_____ United States	_____ Uruguay		

4. Place a "+" on the blank to identify major crops grown on large farms in the mid-latitudes.

_____ apples	_____ bananas	_____ coffee	_____ cacao
_____ corn	_____ cotton	_____ grapefruit	_____ oats
_____ oranges	_____ pears	_____ rice	_____ soybeans
_____ tobacco	_____ wheat		

Using **Map 15a** and **an atlas**, complete the following.

1. The Great Plains region is a major wheat- and cattle-producing region of the world. Connect the dashes to enclose this major wheat- and cattle-producing region.

2. In the Corn Belt, large amounts of corn, soybeans, hogs, and cattle are produced. Connect the dots to enclose the Corn Belt region.

3. In the Great Central Valley of California, citrus fruits and nuts are raised in large quantities. This region is inside the symbol ◁▧▷ . Shade in the area marked by the symbol to show the location of the Great Central Valley.

4. The Willamette Valley in the northwestern United States is an important farming area. The symbol ⊞ marks the Willamette Valley. The Willamette Valley is located in   a) Oregon   b) California c) Texas d) Arkansas.

Name: _____ Date: _____

Unit 7: Agriculture in the Western Hemisphere

B. Mid-Latitude Crops (cont.)

5. The Ontario Peninsula is noted for the production of fruits like peaches and apples. The symbol ▦▦▦ locates the Ontario Peninsula. The peninsula is located in

   a) the United States   b) Canada   c) Brazil.

6. The nearby Great Lakes make the climate on the Ontario Peninsula much milder, so the production of fruits is possible. The Great Lakes that border the Ontario Peninsula are

   a) Lake Michigan and Lake Huron   b) Lake Erie and Lake Michigan

   c) Lake Huron, Lake Ontario, and Lake Erie.

7. Draw a line to connect the X's that indicate the southeastern United States. The southeastern coast of the United States is a region of long, hot summers and cool, mild winters. Many crops are raised. Cotton, tobacco, rice, peanuts, citrus fruits, and vegetables are major crops. Place a + on the blank if the state is located in the southeastern United States.

   _____ Alabama   _____ California   _____ Florida   _____ Georgia

   _____ North Carolina   _____ Louisiana   _____ Maine   _____ Mississippi

   _____ New Hampshire _____ Ohio   _____ South Carolina

Using **Map 15b** and **an atlas**, complete the following. You will need a brown and a blue colored pen or pencil for this activity.

1. The Pampas region is a major wheat-, corn-, and cattle-producing region in South America. Connect the dashes to enclose this major grain-producing area of South America.

2. The Pampas are located in   a) Chile   b) Argentina   c) Bolivia.

3. Patagonia is a plateau region noted for its sheep production. The Patagonia Plateau is located with the symbol ◁▭▭▭▷ . The Patagonia Plateau is located in

   a) California   b) Uruguay   c) Argentina.

4. The Vale of Chile is much like the Great Central Valley in California. Many of the fruits and vegetables produced there are shipped to the United States. The Vale of Chile is located with the symbol ▤▤▤ . The Vale of Chile is located in a) Chile b) Venezuela   c) Brazil.

5. Along the coast of Brazil, many crops are raised. The coastal area is outlined with dashed lines. Color the area along the coast of Brazil brown.

6. Inland cattle production is very important. Connect the dots to locate this region. Color the region blue.

7. The southern part of Brazil is an important agricultural region noted for soybean production. Connect the x's to outline this region.

Name: _____ Date: _____

Unit 7: Agriculture in the Western Hemisphere

B. Mid-Latitude Crops (cont.)

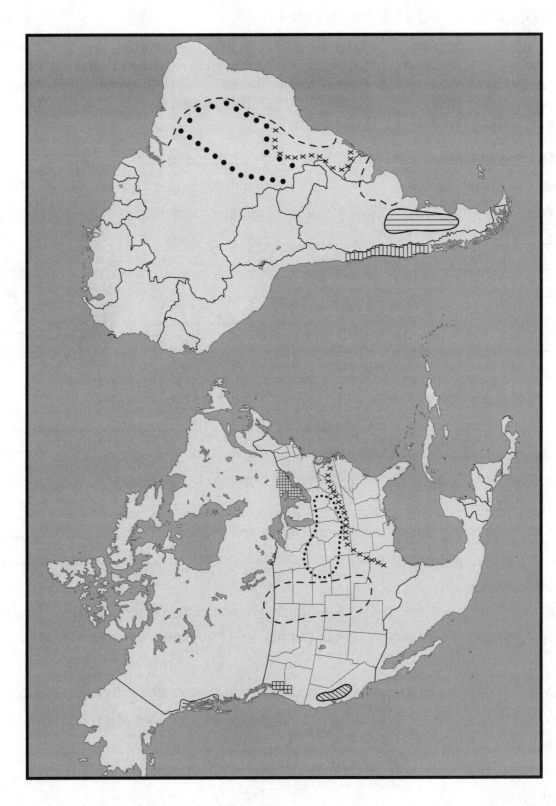

MAP 15b

MAP 15a

Name: _____ Date: _____

Unit 7: Agriculture in the Western Hemisphere

B. Mid-Latitude Crops—Pretest Practice

Complete the blanks using the following terms.

30	60	apples	Argentina	beef	butter	Canada
cheese		chicken	Chile	corn	Corn Belt	cotton
Great Central Valley			Great Plains	milk	Ontario	oranges
Pampas		pears	pork	United States		Uruguay
Vale of Chile		wheat	Willamette			

1. The mid-latitudes are between _____ and _____ degrees north and south latitude.

2. Crops grown in the mid-latitudes include crops like

   _____, _____,

   _____, _____,

   _____, and _____.

3. Animals are raised for products like _____,

   _____, _____,

   _____, and _____.

4. The countries located in the mid-latitudes that produce crops include

   _____, _____, _____,

   _____, and _____.

5. An important citrus fruit- and nut-producing region in California is the

   _____.

6. An important citrus fruit- and nut-producing region in Chile is the _____.

7. An important fruit-producing region in Canada is on the _____ Peninsula.

8. An important wheat- and livestock-producing region in the United States is found in Kansas, Nebraska, North Dakota, South Dakota, and Oklahoma and is known as the

   _____.

9. An important grain- and livestock-producing region in Argentina is the _____.

10. An important grain- and livestock-producing region in the states of Iowa, Illinois, Missouri, Ohio, and Indiana is the _____.

11. An important farming area in the state of Oregon is found in the _____ Valley.

Name: _____ Date: _____

Unit 7: Agriculture in the Western Hemisphere

B. Mid-Latitude Crops—Test

Circle the letter of the correct answer.

1. Mid-latitude crops are grown between 30 degrees and   a) 60   b) 90
    c) 40   d) 100   degrees north and south latitude.

2. Corn, apples, oranges, pears, wheat, and cotton are examples of crops grown in the
    a) high latitudes   b) tropics   c) mid-latitudes.

3. Countries located in the mid-latitudes that produce crops include countries like
    a) Argentina, Honduras, Jamaica   b) the United States, El Salvador, Haiti
    c) Uruguay, Argentina, the United States, Canada   d) Cuba, Trinidad, Brazil.

4. An important citrus fruit- and nut-producing region in California is the   a) Great Plains
    b) Great Central Valley   c) Vale of Chile   d) Corn Belt.

5. An important citrus fruit- and nut-producing region in Chile is the   a) Great Plains
    b) Great Central Valley   c) Vale of Chile   d) Corn Belt.

6. An important fruit-producing region in Canada is the   a) Gaspe   b) Chesapeake
    c) Cape Cod   d) Ontario   Peninsula.

7. An important grain- and cattle-producing region in Argentina is the   a) Pampas
    b) Great Plains   c) Altiplano   d) Great Central Valley.

8. An important farming region in Oregon is the   a) Willamette Valley   b) Great Plains
    c) Corn Belt   d) Great Central Valley.

9. An important wheat- and livestock-producing region in the United States found in Kansas,
    Nebraska, North Dakota, South Dakota, and Oklahoma is known as the
    a) Pampas   b) Willamette Valley   c) Great Plains   d) Corn Belt.

10. An important grain- and livestock-producing region in the states of Iowa, Illinois, Missouri,
    Ohio, and Indiana is the   a) Willamette Valley   b) Great Plains   c) Corn Belt
    d) Great Central Valley.

Unit 8: Natural Resources of the Western Hemisphere

Using **Maps 16**, **17**, and **an atlas**, complete the following.

A. Oil

Oil, iron, copper, bauxite, coal, and forests are natural resources. These natural resources are found in various countries in North America, South America, and the West Indies (Caribbean).

Oil is a very valuable natural resource, and many things are made from it. The gas and oil used by airplanes, automobiles, and trucks are made from petroleum. The plastics that are used for so many purposes in daily life are a petroleum product.

1. The symbol ⚒ locates major oil-producing regions. Place a "+" on the blank if the nation is a major oil-producing nation.

   _____ Canada   _____ Chile   _____ Cuba   _____ Mexico

   _____ Trinidad   _____ United States   _____ Venezuela

2. Two states in the United States that are important oil producers are

   a) Illinois and Iowa   b) Florida and North Carolina   c) Texas and Alaska

   d) Tennessee and Ohio.

3. An important oil-producing region in Canada is found in   a) Alberta

   b) Labrador   c) Nova Scotia   d) Prince Edward Island.

4. In Mexico, an important oil-producing region is   a) Baja California b) Yucatán Peninsula.

B. Iron Ore

Nations with large **iron ore** deposits are usually important steel-producing nations. Large iron ore deposits are found in North America and South America. On Maps 16 and 17, iron ore deposits are located with the symbol ⚒ .

5. Place a "+" on the blank if the nation has large iron ore deposits.

   _____ Brazil   _____ Canada   _____ Ecuador   _____ El Salvador

   _____ Dominican Republic   _____ Venezuela

6. An important iron ore-producing region in Canada is found in

   a) Edmonton   b) Labrador   c) Nova Scotia   d) Prince Edward Island.

7. In South America, large iron ore deposits are found in   a) Chile and Peru

   b) Venezuela and Brazil   c) Mexico and Belize   d) Colombia and Panama.

Name: _____ Date: _____

Unit 8: Natural Resources of the Western Hemisphere

C. Copper

Copper mines are found in both North America and South America. Copper is an important mineral. It is a very good conductor, so electrical wire is often made from copper. It is also used with tin to make bronze. Copper deposits are indicated on Maps 16 and 17 by the chemical symbol **Cu**.

8. Large copper deposits in South America are found in   a) Chile   b) Venezuela
   c) Paraguay.

9. Large copper deposits in North America are found in   a) the United States and Canada
   b) Canada and Mexico   c) Mexico and the United States.

10. In the United States, the states of   a) California, Oregon, and Idaho   b) Utah, Arizona, and New Mexico   c) West Virginia and Maryland   have large copper deposits.

D. Bauxite

Bauxite is the mineral from which aluminum is made. Bauxite deposits are indicated on Maps 16 and 17 by the chemical symbol **Al**.

11. Large bauxite deposits are found on the West Indies island nation of a) Trinidad
    b) Cuba c) Jamaica.

12. In South America, large deposits of bauxite are found in   a) Brazil and Argentina
    b) Chile and Peru c) Guyana and Suriname.

E. Coal

Large **coal** deposits are found in North America. Coal types include lignite, bituminous, and anthracite. Bituminous deposits are more common and more extensive and provide fuel for many industries. The United States has large deposits of bituminous coal. These large bituminous coal deposits have been very important in the industrial development of the nation. Bituminous coal deposits are located on Map 16 using the symbol **C**.

13. Large deposits of bituminous coal are found in   a) Texas and Utah b) Illinois and Pennsylvania c) Mexico and Canada.

F. Forests

Large **forests** are found in both North America and South America. Today, forests are not nearly as numerous as they once were. The large forests make the lumber industry very important. Maps 16 and 17 show the location of the forested areas using the tree symbol 🌲 .

14. In South America, large forests are found in   a) Cuba and Panama
    b) Uruguay and Argentina   c) Brazil and Trinidad   d) Chile and Brazil.

15. In Canada, large forests are found on the Canadian Shield in   a) Saskatchewan
    b) Prince Edward Island   c) Yukon Territory   d) Ontario and Quebec.

16. In Canada, large forests are found in the Rocky Mountains in   a) British Columbia
    b) Nova Scotia   c) Prince Edward Island   d) Labrador.

17. In the United States, large forests are found in   a) Texas and New Mexico
    b) Kansas and Iowa   c) Mississippi and Florida   d) Oregon and Washington.

Unit 8: Natural Resources of the Western Hemisphere

MAP 17

MAP 16

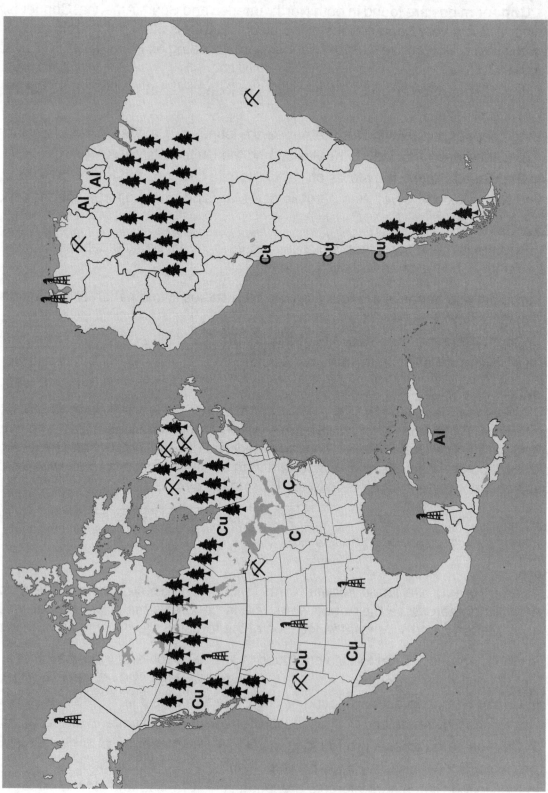

Name: _____   Date: _____

Unit 8: Natural Resources of the Western Hemisphere— Pretest Practice

Use the terms below to complete the blanks.

Bingham Canyon	**Canada**	**Canadian Shield**	**Chile**	**coal**
copper	**El Pao**	**Jamaica**	**Labrador**	**lumber**
Maracaibo Basin	**Mesabi Range**	**Minas Gerais**	**oil**	**Trinidad**
United States	**Venezuela**			

1. This South American nation produces large amounts of oil. Most of the oil is shipped to the United States. The nation is _____.

2. An island nation near Venezuela that is a major oil producer is _____.

3. A major oil-producing region in Venezuela is the region around Lake Maracaibo known as the _____.

4. Large deposits of iron ore are found in northeastern Canada in _____.

5. Large deposits of copper are found in the Atacama Desert in northern _____.

6. A state in Brazil where large deposits of iron ore are found is _____.

7. In the United States, the _____ was once an important iron ore-producing region in Minnesota.

8. The _____ iron ore deposits are found in Venezuela.

9. In the United States, a large copper mine is located in Utah. It is known as the _____ mine.

10. This large shield-shaped region of very hard rock that was left when glaciers stripped the soil away is located in Canada. The region is now noted for its forests and lumber industry. Where rivers leave the shield and tumble down to the St. Lawrence River, hydroelectric plants are found. The large shield-shaped area of rock is the _____ _____.

11. Mexico is an important producer of _____, which is shipped to the United States.

12. Large deposits of _____ are found in Chile.

13. Large deposits of bauxite used to produce aluminum are found in _____.

14. Large deposits of bituminous _____ are found in the United States.

15. The forests of the Amazon Basin in Brazil produce large amounts of _____.

16. A North American country that has large forests and produces lumber is _____.

81

Unit 8: Natural Resources of the Western Hemisphere—Test

Circle the letter of the correct answer.

1. This South American nation produces large amounts of oil. Most of the oil is shipped to the United States. The nation is   a) Chile   b) Argentina   c) Venezuela   d) Uruguay.

2. An island nation near Venezuela that is a major oil producer is   a) Trinidad   b) Cuba   c) Galapagos   d) Jamaica.

3. A major oil-producing region in Venezuela is the region around Lake Maracaibo known as the   a) Vale of Chile   b) Pampas   c) Llanos   d) Maracaibo Basin.

4. Large deposits of iron ore are found in northeastern Canada in   a) Labrador   b) Yukon Territory   c) the Northwest Territories   d) Alberta.

5. Large deposits of copper are found in the Altacama Desert in northern   a) Colombia   b) Cuba   c) Chile   d) Argentina.

6. A state in Brazil where large deposits of iron ore are found is   a) Minas Gerais   b) Mato Grosso   c) Paraná   d) Amazonas.

7. The region in Minnesota that was once an important iron ore-producing region is the   a) El Pao   b) Bingham Canyon   c) Mesabi Range.

8. The iron ore deposits found in Venezuela are the   a) El Pao   b) Bingham Canyon   c) Mesabi Range   deposits.

9. The large copper mine located in Utah is the   a) El Pao   b) Bingham Canyon   c) Mesabi Range   mine.

10. The Canadian Shield is a large shield-shaped region important for   a) oil   b) large forests and the lumber industry   c) farming.

11. Mexico is an important producer of   a) oil   b) lumber   c) coal   which is shipped to the United States.

12. Large deposits of   a) copper   b) tin   c) oil   d) coal   are found in Chile.

13. Large deposits of bauxite used to produce aluminum are found in   a) Chile   b) Jamaica   c) Uruguay   d) Colombia.

14. Large deposits of coal are found in   a) the United States   b) Uruguay   c) Argentina   d) Jamaica.

15. The forests of the Amazon Basin in Brazil produce large amounts of   a) lumber   b) tin   c) oil   d) coal.

16. A country that has large forests and produces lumber is   a) Uruguay   b) Argentina   c) Canada   d) Cuba.

Name: _____ Date: _____

Unit 9: Central America and the West Indies

A. Central America

Central America is an isthmus. An isthmus is a neck of land that connects two larger landmasses. In the case of Central America, it is an isthmus that connects North and South America.

1. Use **an atlas** and **Map 18**, which shows the Central American isthmus, for this activity. Central America includes seven countries located on the isthmus. Place a "+" on the blank if the country is a Central American country on the isthmus.

   _____ Belize   _____ Canada   _____ Chile   _____ Costa Rica

   _____ Cuba   _____ El Salvador   _____ Guatemala   _____ Honduras

   _____ Nicaragua   _____ Panama   _____ Trinidad

In Central America, three main climates are tropical rain forest, tropical savanna, and highland in the mountains. The tropical rain forest and tropical savanna both have year-round high temperatures and lots of rain. The tropical rain forest has rain in every month without a dry season. The vegetation consists of trees, which are home to many animal varieties. The tropical savanna has much rain in summer, but winters are dry. The vegetation is usually shrubs and tall grass. The highland climate is much cooler and milder than the very warm temperatures and rainy climate of the tropical rain forest and tropical savanna.

Central American countries are all in the tropics. Except in the mountains, the tropics are a region of year-round warm temperatures. Those regions with abundant rain year-round have winds that blow toward the shore throughout the year. The winds blow onshore and rise over the mountains, producing abundant rain. Those regions with a wet season followed by a dry season have winds blowing from over the water toward the shore in the wet season. In the dry season, the winds blow from over the land toward the water.

MAP 18

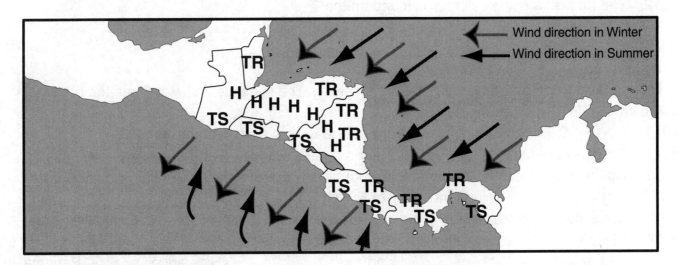

Name: _____ Date: _____

Unit 9: Central America and the West Indies

A. Central America (cont.)

Using **Map 18**, complete the following. The arrows show wind direction.

2. Along the eastern coast of Central America, the winds blow from over the water
   a) in summer and winter   b) in summer and from land to water in winter.
3. Along the eastern coast of Central America, the temperatures would be warm all year and there would be   a) a definite wet and dry season   b) rainfall throughout the year.
4. Along the western coast of Central America, the winds blow from over the water   a) all year long   b) in summer and from land to water in winter.
5. Along the western coast of Central America, the temperatures would be warm all year and there would be   a) a definite wet and dry season   b) rainfall throughout the year.

Using **Map 18** and **an atlas**, complete the following. You will need red, pink, and black colored pens or pencils for this activity. The areas with the letters "TR" have a tropical rain forest climate. The areas with the letters "TS" have a tropical savanna climate. The areas with the letter "H" have a highland climate.

6. Color the tropical rain forest climate red.
7. Color the tropical savanna climate pink.
8. Color the highland climate black.
9. Place a "+" on the blank if the country has a highland climate.
   _____ Belize   _____ Costa Rica   _____ El Salvador   _____ Guatemala
   _____ Honduras   _____ Nicaragua   _____ Panama
10. The tropical rain forest climate is found on the
    a) west coast   b) east coast   of the Central American countries.
11. Place a "+" on the blank if the country has some areas of tropical rain forest climate.
    _____ Belize   _____ Costa Rica   _____ El Salvador   _____ Guatemala
    _____ Honduras   _____ Nicaragua   _____ Panama
12. The tropical savanna climate is found on the
    a) west coast   b) east coast   of the Central American countries.

Agriculture

Using **Map 19** and **an atlas**, complete the following. The symbols below indicate the areas for the crops raised in Central America.

Cs = coffee   **s = sugar cane**   **b = bananas**

1. Coffee is produced in the   a) highlands   b) lowlands   of Central America.
2. Coffee is produced in all countries except   a) Panama and Belize
   b) Nicaragua and Honduras   c) El Salvador and Costa Rica.
3. Sugar cane is produced along the   a) west coast   b) east coast   of Central America.
4. Place a "+" on the blank if the country produces bananas.
   _____ Costa Rica   _____ Honduras   _____ Nicaragua _____ Panama

Unit 9: Central America and the West Indies

B. The West Indies (Caribbean Islands)

The **West Indies** is the name given to the tropical islands in the Caribbean. There are many islands in the West Indies, and most of them are very small.

1. Using **Map 19** and **an atlas**, complete the following about the West Indies. Place the letter by the country on the map to show its location.

   West Indies (Caribbean) countries
   A. Cuba   B. Jamaica   C. Haiti   D. Dominican Republic   E. Puerto Rico

2. A dot on **Map 19** indicates the capital of each Caribbean island. Place the letter of the capital city next to the corresponding dot on the map.

   Capital Cities
   a. Havana   b. Kingston   c. Port-au-Prince   d. Santo Domingo   e. San Juan

Climate

The most common climate type in the West Indies is the tropical savanna.
The climate chart below shows the average monthly temperature and rainfall typically found for locations in the West Indies. Use the chart to complete the questions that follow.

Month	J	F	M	A	M	J	J	A	S	O	N	D	Year
Temperature (°F)	76	77	77	78	79	81	82	82	81	80	79	77	79
Rainfall (inches)	1.0	1.0	1.0	1.5	4.0	4.0	3.5	3.5	4.0	3.0	1.0	1.0	28.5

1. The yearly temperature range is   a) 3   b) 6   c) 10   d) 12   degrees.
2. The rainy season extends from   a) May to August   b) May to October   c) January to May.
3. True or False. It is warm throughout the year in the West Indies.
4. True or False. There is a definite dry season in the West Indies.

Agriculture

Major crops grown in the West Indies are located using the following symbols.
   Cs = coffee   s = sugar cane   t = tobacco   b = bananas   CB = cacao beans

Using **Map 19** and **an atlas**, circle the letters of the correct answers.
1. Two nations in the West Indies that produce large amounts of sugar cane are   a) Cuba and the Dominican Republic   b) Brazil and Cuba   c) Honduras and Venezuela.
2. Two nations in the West Indies that produce cacao beans are   a) Cuba and Belize   b) Chile and Puerto Rico   c) Haiti and the Dominican Republic.
3. Two nations in the West Indies that produce coffee are   a) Aruba and Costa Rica   b) Puerto Rico and Haiti   c) Colombia and Panama.
4. Two nations in the West Indies that produce tobacco are   a) Aruba and Trinidad   b) Cuba and the Dominican Republic   c) Jamaica and Puerto Rico.
5. Two nations in the West Indies that produce bananas are   a) Jamaica and Haiti   b) Panama and Trinidad   c) Puerto Rico and Peru.

Unit 9: Central America and the West Indies

B. The West Indies (Caribbean Islands) (cont.)

MAP 19

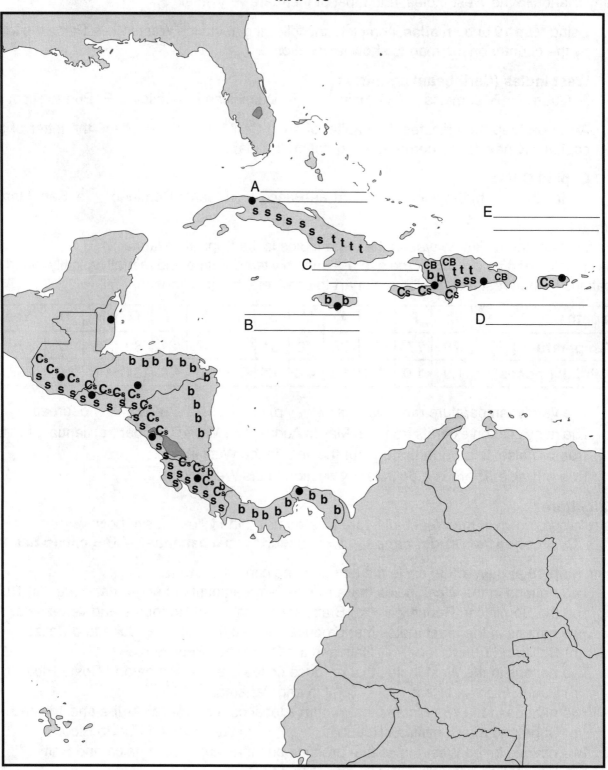

Name: _____   Date: _____

Unit 9: Central America and the West Indies

C. Pretest Practice

Central America: Complete the following blanks using the terms below.

offshore	coffee	dry	east
isthmus	bananas	onshore	tropics
sugar cane	warm	west	wet

1. The Central American countries are all located on the _____ connecting North America to South America.
2. In the Central American countries, the temperature is _____ all year.
3. The tropical rain forest climate is located on the _____ coast of Central America.
4. On the east coast of Central America, the winds blow _____ all year.
5. The tropical rain forest climate is warm and _____ every month of the year.
6. The tropical savanna climate is located on the _____ coast of Central America.
7. On the west coast of Central America, the winds blow onshore part of the year but _____ the other part of the year.
8. The west coast of Central America has a definite _____ and _____ season.
9. The wet season on the west coast comes when the winds are blowing _____.
10. The dry season on the west coast comes when the winds are blowing _____.
11. Important crops raised on large plantations in Central America include _____, _____, and _____.
12. A crop raised in the highlands and shipped to the United States is _____.

The West Indies: Complete the blanks below using the following terms.

bananas	cacao	coffee	sugar cane	summer
temperature	tobacco	tropical savanna		

1. Cuba is an important producer of _____.
2. Haiti and the Dominican Republic are important producers of _____.
3. Puerto Rico and Haiti are producers of _____.
4. Jamaica and Haiti produce _____.
5. Cuba and the Dominican Republic produce _____.
6. The climate most common in the West Indies is the _____.
7. The rainy season in the West Indies comes during the _____ and early fall.
8. There is little difference in _____ between the warmest and coolest months in the West Indies.

Name: _____ Date: _____

Unit 9: Central America and the West Indies

C. Test

Circle the letter of the correct answer.

1. The Central American countries are all located in the   a) mid-latitudes   b) polar latitudes   c) tropic latitudes.
2. In the Central American countries, the temperature is   a) warm   b) cool   c) cold throughout the year.
3. The tropical rain forest climate is located on the   a) west   b) south   c) east   d) north   coast of Central America.
4. On the east coast of Central America, the winds blow   a) offshore   b) onshore   all year.
5. The tropical rain forest climate is warm and   a) wet   b) dry   every month of the year.
6. The tropical savanna climate is located on the   a) west   b) south   c) east   d) north   coast of Central America.
7. On the west coast of Central America, the winds blow onshore part of the year but   a) offshore   b) onshore   the other part of the year.
8. The west coast of Central America has   a) only a wet   b) only a dry   c) a wet and a dry   season.
9. The wet season on the west coast comes when the winds are blowing a) onshore   b) offshore.
10. The dry season on the west coast comes when the winds are blowing   a) onshore   b) offshore.
11. Important crops raised on large plantations in Central America include   a) wheat and corn   b) coffee and bananas   c) bananas and soybeans.
12. An important crop raised in the highlands is   a) bananas   b) wheat   c) cotton   d) coffee.
13. Sugar cane production is very important in the West Indies country of a) Cuba   b) Brazil   c) San Salvador   d) Costa Rica.
14. Two West Indies countries that are important producers of cacao are   a) Chile and Trinidad   b) Puerto Rico and Colombia   c) Haiti and the Dominican Republic   d) Jamaica and Costa Rica.
15. Two West Indies countries that are important producers of coffee are   a) Puerto Rico and Haiti   b) Puerto Rico and Belize   c) Haiti and Mexico   d) Trinidad and Panama.
16. Two West Indies countries that are important producers of bananas are   a) El Salvador and Jamaica   b) Cuba and Venezuela   c) Aruba and Costa Rica   d) Jamaica and Haiti.
17. Two West Indies countries that are important producers of tobacco are   a) Aruba and Brazil   b) Trinidad and Belize   c) Cuba and the Dominican Republic   d) Puerto Rico and Honduras.
18. The climate most common in the West Indies is the   a) tropical rain forest   b) tropical savanna   c) highland   d) desert.
19. The rainy season in the West Indies comes during the   a) summer and early fall   b) winter and spring   c) fall and winter   d) spring and summer.
20. In the West Indies, the average temperature difference between the warmest and coldest month is nearest to   a) 40   b) 6   c) 20   d) 18 degrees.

Discovering the World of Geography: Grades 6–7   Unit 10: Which Country, State, Province, or Territory?

Name: _____   Date: _____

Unit 10: Which Country, State, Province, or Territory Is Described?

A.   The statements below identify a country, state, province, or territory in the Western Hemisphere. Use wall maps, an atlas, or maps from this book that show political boundaries, physical features, and climate information for the Western Hemisphere nations, and then complete the following.

1.   Located in North America, this country was the home of the Aztec Indian civilization. This country shares a 2,500-mile border with its neighbor to the north. The Rio Grande is a boundary between the country and its northern neighbor. The northern part of the nation is desert, but farther south it becomes mountainous. The capital is a very large city located at an elevation of over 7,000 feet. From the capital, the snow-covered volcanic peaks of Mt. Popocatepetl and Mt. Iztaccihuatl can be seen in the distance.

The country is a) _____, which is located in the b) (Northern / Southern) Hemisphere. Its neighbor to the north is c) _____.

2.   A large part of the population of this South American country is mestizo. When Spaniards came to South America, they often married Indian women; the children with a European father and Indian mother are known as *mestizos*. This country has large deposits of oil and iron. Much of the oil is shipped to the United States. Its coastal areas on the Caribbean Sea are warm and humid. So the people could escape the heat and humidity of the coast, the capital was built in a mountain valley where the climate is much cooler. A large river is the Orinoco, which drains a large plains area known as the Orinoco Plain. Lake Maracaibo, located along the Caribbean coast, has large deposits of oil.

The country is a) _____, which is located in the b) (Northern / Southern) Hemisphere. The largest country in South America, located on the southern border, is

c) _____.

3.   This country is located in the West Indies. It is one of two countries located on a large island. This country is located on the eastern side of the island. Sugar cane is an important crop that is shipped to many other nations. The language spoken in this country is Spanish, while citizens of the other country on the island speak French. Puerto Rico is a neighbor to the east.

Discovering the World of Geography: Grades 6–7   Unit 10: Which Country, State, Province, or Territory?

Name: _____   Date: _____

Unit 10: Which Country, State, Province, or Territory Is Described? (cont.)

The country is a) _____, which is located on the island of b) _____. The other country located on the island is c) _____.

4. This state is located in the Western Hemisphere. In population, it is the largest state in the United States. Between 30° and 40° north latitude, the state has a Mediterranean climate. On the eastern border, the Sierra Nevadas are found with Mt. Whitney towering 14,494 feet. The Mojave Desert, Death Valley, and the Salton Sea are located in this state. In the north, Mt. Shasta and the redwood forests are famous. President Nixon was born in this state.

   The state is a) _____. On the western boundary is the b) _____ Ocean. Two large cities in the southern part of the state are c) _____ and d) _____.

5. This province is located in a large country in the Northern Hemisphere. The French and English have been important in the history of this province. It is one of two peninsula provinces located in this country. The Bay of Fundy, with its extremely high and low tides, separates this province from the other peninsula province. Although the province is located at a high latitude, farming is very important. Since it is located near the Grand Banks in the Atlantic Ocean, fishing is one of the main industries. It is the historical home of the French Acadians, many of whom left the region in 1775 and settled near New Orleans, Louisiana, which was controlled by Spain. In Louisiana, the Acadians became known as "Cajuns."

   The province is in the country of a) _____. The province is b) _____. The capital city is c) _____. The Bay of Fundy separates this province from the province of d) _____.

6. The Colorado River flows through this state. Over time, the Colorado River formed the Grand Canyon, which is located in the state. Much of the state has a semiarid or desert climate. The southern boundary of the state forms part of the border with Mexico.

   The state is a) _____ in the country of b) _____. The capital city is c) _____.

Name: _____ Date: _____

Unit 10: Which Country, State, Province, or Territory Is Described? (cont.)

7. This country is located in South America. While it is over 2,600 miles long, it is only 200 miles wide. In the north is the Atacama, one of the driest deserts in the world. Copper mines are found in the northern part of the country. In the central part of the country, the climate is Mediterranean, much like California. Citrus fruits and vegetables are grown on the very productive land in this region. Near the southern tip of the country is Cape Horn. Just offshore is the cold Peru Current. Mt. Aconcagua, one of the highest peaks in South America, is located just across the eastern border in a neighboring country.

    The country is a) _____. The capital city is b) _____. The

    c) _____ Mountains separate the country from its neighbor to the east. The

    country is located on the continent of d) _____. The neighbor to the east is

    e) _____.

8. This is the largest country in South America. In most South American countries, Spanish is spoken, but in this country, Portuguese is spoken. The Amazon River is found in this country. The production of crops like soybeans and corn is very important in the southern part of the country. Most cities are located on the coast, but the capital is located inland. The country has large iron ore deposits. The tropical rain forest and tropical savanna climates are found in this country.

    The country is a) _____. The capital city is b) _____.

9. This small country is located on the isthmus that connects North America and South America. It faces the Pacific Ocean on the west and a much larger neighboring country on the east. An important crop is coffee. Like other nearby countries, Spanish is the national language. During the summer, the warm winds from over the ocean blow onshore, bringing a lot of rain. In the winter, the winds blow offshore toward the sea, and there is a very dry season.

    The country is a) _____. The capital city is b) _____. This coun-

    try, on the isthmus connecting North and South America, is part of c) _____

    America.

10. This country is one of two landlocked countries in South America. In this country, the quebracho trees are found. The wood from these trees is very hard. The bark of the tree is used in the tanning of leather. A large part of this country is in the plains region called the Gran Chaco. The Gran Chaco is where the quebracho tree is found.

    The country is a) _____. The capital city is b) _____. The re-

    gion where the quebracho tree grows is the c) _____.

Name: _____ Date: _____

Unit 10: Which Country, State, Province, or Territory Is Described? (cont.)

B. To complete the following exercise, an atlas, wall map, or a map from this book showing political boundaries, physical features, latitude and longitude, and climate types will be needed. The statements below describe a country, state, or province. Use the statements to complete the blanks.

1. Located high in the Andes Mountains, this country is landlocked. The country has two capitals located on the high Altiplano, which is a high plateau surrounded by the Andes Mountains. The Altiplano was also home to the Inca Indian civilization that was found there when the Spanish explorers came to South America. Rich deposits of copper and antimony are found here. Two large lakes are found on the Altiplano. In the high Andes, overlooking the Altiplano, is snow-capped Mt. Illimani.

   The country is a) _____. The capital cities are b) _____ and

   c) _____. The large lakes are Lake d) _____ and Lake

   e) _____.

2. This island province is the smallest in the country in which it is located. The capital is located at 46 degrees north latitude and 63 degrees west longitude. The climate is much milder than would be expected, because the province is protected from the Atlantic storms by the province of Nova Scotia. Potatoes and tobacco are important crops.

   The province is a) _____. The capital city is b) _____.
   The province is located in the country of c) _____. The large gulf near the
   province is the d) _____.

3. This country is located between 20 degrees and 23 $\frac{1}{2}$ degrees north latitude. To the west is a large gulf. To the east is the Atlantic Ocean. To the south is a large tropical sea. To the north 90 miles is a large country. The most important crop is sugar cane. It is the largest country in the West Indies.

   The country is a) _____. The capital city is b) _____. The large
   gulf to the west is the c) _____. The tropical sea to the south
   is the d) _____.

4. This country is located at 80 degrees west longitude. The capital city located high in the mountains is $\frac{1}{2}$ degree south of the equator. From the capital, the snow-capped volcano Mt. Cotopaxi can be seen. Directly west in the Pacific Ocean is a famous island chain known for its large tortoises. Along the Pacific coast, the major port city is located.

   The country is a) _____. The capital city is b) _____. The famous islands to the west are the c) _____. A port city in this country is

   d) _____.

Name: _____   Date: _____

Unit 10: Which Country, State, Province, or Territory Is Described? (cont.)

5. Near the coast, these states have very mild winters and summers. There are many cloudy days with light rainfall. Near the coast is a high mountain chain with large forests. The lumber industry is very important to these states, as is salmon fishing in the rivers. The fertile Willamette Valley is located in one of the states. In one state, Mt. Rainier is located, while in the other, Mt. Hood is located. Apple production in the Yakima Valley is important. Wheat is grown in the Palouse Hill region.

The states are a) _____ and b) _____. The capital cities are c) _____ and d) _____. A major river that empties into the Pacific Ocean is the e) _____. The high mountain chain is the f) _____.

6. The Mississippi River is the eastern boundary of these two states. Directly to the north is the Hawkeye State and to the south the Bayou State. President Truman was born in one state and President Clinton in the other. The Ozark Plateau is found in both states. The Ozark Plateau is a tourist region of hilly, scenic land and has many large lakes, small rivers, and streams. The Boston Mountains are located in the southernmost state. In one state, a city is known as the Gateway to the West. This city is located near the point where the Missouri River joins the Mississippi River and where Lewis and Clark began their exploration of the far west. Samuel Clemens, later known as Mark Twain, grew up in one of these states, in the town of Hannibal on the Mississippi River.

The states are a) _____ and b) _____. The capital cities are c) _____ and d) _____. The country in which the states are located is e) _____. The city that is the "Gateway to the West" is f) _____. The Boston Mountains are located in the state of g) _____. The "Hawkeye State" is the state of h) _____. The "Bayou State" is the state of i) _____.

Name: _____ Date: _____

Unit 10: Which Country, State, Province, or Territory Is Described: Crossword Puzzle—Canada

Use an atlas or globe to complete the following crossword puzzle.

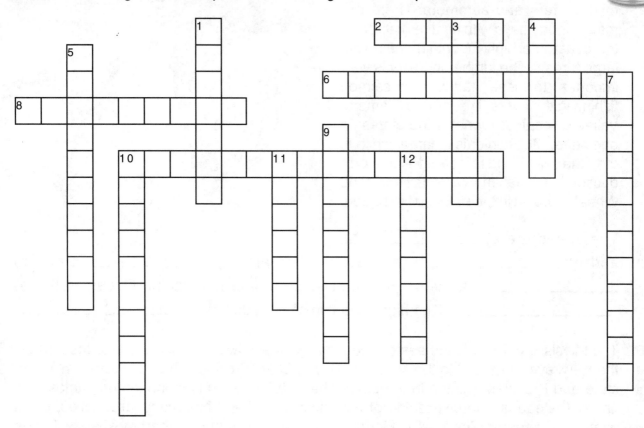

ACROSS

2. Territory that borders Alaska
6. This Canadian province borders the states of Montana and North Dakota. This province is part of the Great Plains and produces wheat.
8. This large island is located in the Pacific Ocean. It is part of the westernmost province in Canada. Victoria is the province's capital city.
10. The Rocky Mountains are located in this province. Along the Pacific coast, the climate is mild, with cloudy, rainy days. The lumber industry is important. (two words)

DOWN

1. This province is important for wheat production. It also has large deposits of oil. On the south, it borders Montana.
3. This province is the only Canadian province with a border on the Great Lakes.

4. Large parts of this province are on the Canadian Shield. The lumber industry is very important. French is the language spoken here.
5. In the subarctic climate regions of Canada, there are large forests of pines, firs, and spruce. These evergreens are also known as _____ trees.
7. This province includes an island. Labrador on the mainland is part of this province. Iron ore deposits are found here.
9. Living in the far north of Canada near the Arctic Circle are the Inuits, or Eskimos. In 1999, the far eastern part of this territory became Nunavut Territory, which was once a part of the _____ Territories.
10. This bay is located between New Brunswick and Nova Scotia and is known for extremely high and low tides. (three words)
11. This large bay is found in far-northern Canada.
12. Lake Winnipeg is located in this province.

Name: _____ Date: _____

Use an atlas or globe to complete the following crossword puzzle.

ACROSS

1. This is the only Great Lake that does not border Canada. (two words)
8. Yellowstone National Park is located in this state.
9. Four states in the Southwest are known as the "Four Corners States," because the borders of the four states meet at the same point. These states are Arizona, New Mexico, Colorado, and _____.
11. The Colorado River formed the Grand Canyon and flows through the northwest corner of this state.
12. The neighboring state to the west of this state has more people in it than in any other state in the U.S. Lake Mead and the Great Basin are located here.
13. Only Alaska is larger in area than this state. It is noted for oil and natural gas production. Farming and ranching are important. Cattle, cotton, fruits, vegetables, and other farm products are produced. Across the Rio Grande is Mexico.

DOWN

1. The Mississippi delta is located in this state.
2. This state shares the Sault Ste. Marie Canal with Canada. It is known for automobile production.
3. The Salton Sea, Death Valley, and Mt. Whitney are located in this state.
4. The Snake River flows through this state. It is also known for potato farming.
5. The Rio Grande flows through this state before becoming the border between Texas and Mexico. (two words)
6. A large swamp area covered with grass, known as the Everglades, and Lake Okeechobee are located in this state.
7. Wheat farming is very important in this Great Plains state.
10. The Hudson River Valley is located in this state. (two words)

Name: _____ Date: _____

Unit 10: Which Country, State, Province, or Territory Is Described: Crossword Puzzle—Central and South America

Use an atlas or globe to complete the following crossword puzzle.

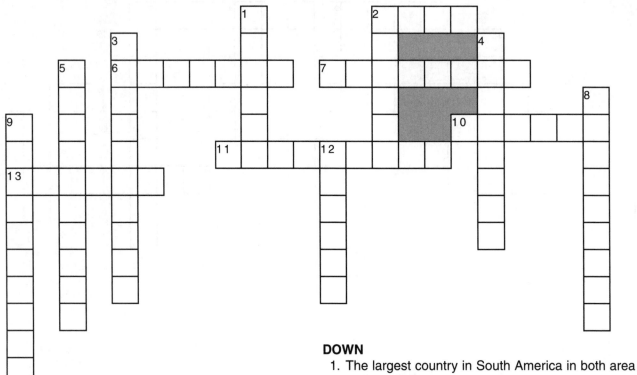

DOWN

1. The largest country in South America in both area and population.
2. This canal connects the Atlantic and Pacific Oceans.
3. This island nation is a commonwealth of the United States. It is located east of Hispaniola in the West Indies. (two words)
4. This cape is located at the tip of South America. (two words)
5. On this island in the West Indies, two countries are found. One is French-speaking; the other is Spanish-speaking.
8. This country is located on the isthmus that connects North and South America. Coffee grown in the highlands and bananas grown along the Caribbean coast are important crops. Most of the population are of European origin. To the south is the country Panama, and to the north is Nicaragua. (two words)
9. This peninsula is part of Mexico. It borders California on the north. The peninsula is called Baja _____.
12. This large, fertile farming region is in Argentina. When Europeans came to Argentina, it was a plains grassland. Today, it is noted for the production of cattle, corn, and other grains. The name of the region is the _____.

ACROSS

2. In this country, near the city of Cuzco, the Inca Civilization flourished.
6. This small country is located between Brazil and Argentina.
7. This country is noted for coffee and banana production and is located on the isthmus between North and South America. The Caribbean Sea is its eastern border.
10. Located in North America, this is a Spanish-speaking country. It was once the home of the Aztec civilization. Across the Rio Grande to the north is its English-speaking neighbor.
11. This high plateau, located in Bolivia, is where most people in the country live.
13. This plains region in Venezuela is a Spanish word that means a large, grassy plain. This plains region is noted for its cattle production. Large deposits of iron are found nearby at Cerro Bolivar. The name given to this plains region is _____.

Name: _____ Date: _____

Unit 11: Where Are You? What Are the Physical Characteristics? What Are the Human Characteristics?

To the Teacher: *National Geography Standard 4:* The human and physical characteristics of places.

Students must become adept at asking questions relating to human and physical characteristics. Students must be encouraged to ask questions relating to the human and physical characteristics of the various stops along the route in this activity. The purpose of this chapter is to help students develop skills that will relate to *National Geography Standard 4*. What are the physical characteristics—mountains, rivers, plains, climate, etc.? What are the human characteristics? What do people do? What about religion, language, and population? Why is it located here? Why is it important? How are they governed? Is the question a human characteristic? A physical characteristic?

Using **Map 20** and **an atlas**, complete the following trip. A dot and the letter "A" mark the beginning point on this trip. A dot and letter locate places where stops will be made. As you take the trip, answer the questions that follow. For each stop made on the trip, you are to write a question to find out more about the human and physical characteristics of the place where the stop is made. For each stop, you must ask a different question and check it as a human characteristic or physical characteristic question.

1. The trip begins at "A," which is located in the country of   a) Venezuela b) Canada
    c) Mexico   d) United States   and the province   a) Manitoba b) Texas c) Mato Grosso   d) Newfoundland.
    Write a question you would like to ask to find out more about this province.
    a. _____
    Human question? _____   Physical question? _____

2. The next stop is at "B," which is a city important in American history. The city is   a) Boston
    b) Regina   c) Miami   d) Philadelphia. The city is located in   a) Canada   b) the United
    States   c) Mexico   d) Cuba.
    Write a question you would like to ask to find out more about this city.
    b. _____
    Human question? _____   Physical question? _____

3. The next city is at "C." To get there, you must sail up the Delaware Bay. The city is known
    as the "City of Brotherly Love." In this city, the Continental Congress met and debated
    whether the Thirteen Colonies should break away from Great Britain and become inde-
    pendent.
    The city is   a) Miami b) Caracas c) Boston d) Philadelphia   located in the state of a)
    Texas b) Pennsylvania c) Delaware d) New York.
    Write a question you would like to ask to find out more about this city.
    c. _____
    Human question? _____   Physical question? _____

Name: _____ Date: _____

Unit 11: Where Are You? What Are the Physical Characteristics? What Are the Human Characteristics? (cont.)

4. The next city visited is located at "D." It is on the Potomac River. This city has many national monuments. The Lincoln Memorial and the Washington Monument are located here. The President of the United States lives in this city. The House of Representatives and the Senate meet here.

   The city is   a) Washington, D.C. b) Philadelphia c) New Orleans d) Boston.

   Write a question you would like to ask to find out more about this city.

   d. _____

   _____

   Human question? _____

   Physical question? _____

5. The next city is located at "E" on a peninsula. Lake Okeechobee and the Everglades are nearby. This city has many Spanish-speaking citizens who came to the city from the nearby island of Cuba.

   The city is   a) Galveston b) Charleston   c) New York d) Miami. This city is located in a) Georgia b) South Carolina c) Florida d) Texas.

   Write a question you would like to ask to find out more about this city.

   e. _____

   Human question? _____   Physical question? _____

6. As you sail south across the Straits of Florida, you reach "F," the largest island in the West Indies. Sugar cane is a very important crop raised on the island. The United States has a large naval base located on the island at Guantánamo.

   The country is   a) Cuba b) the United States   c) Jamaica d) Haiti. The capital city is a) Port-au-Prince   b) Kingston   c) Havana   d) Miami.

   Write a question you would like to ask to find out more about this island.

   f. _____

   Human question? _____   Physical question? _____

7. Continuing on in a southeasterly direction, you reach the next island, "G," which is a large island with two countries. One country speaks French, and the other, Spanish.

   The large island is   a) Trinidad b) Jamaica c) Puerto Rico d) Hispaniola. The two countries located on the island are   a) Cuba and Jamaica b) Haiti and Cayenne c) Haiti and the Dominican Republic d) the Dominican Republic and Trinidad.

Name: _____   Date: _____

Unit 11: Where Are You? What Are the Physical Characteristics? What Are the Human Characteristics? (cont.)

Write a question you would like to ask to find out more about this island.

g. _____

Human question? _____   Physical question? _____

8. Sailing south, you reach the coast of a large oil-producing country, "H." Much of the oil is produced from oil deposits found near a large lake. A large river that runs through the country is the Rio Orinoco, which empties into the Atlantic.

   The capital city is located in the mountains away from the hot, humid coast. The country is a) Colombia b) Cuba c) Venezuela d) Guyana.

   The capital city located in the mountains is a) Caracas b) Santiago c) São Paulo d) Belém.

   The large lake is   a) Lake Michigan b) Lake Poopó c) Lake Titicaca d) Lake Maracaibo.

   Write a question you would like to ask to find out more about this country.

   h. _____

   Human question? _____   Physical question? _____

9. The next city visited, "I," is located along the coast, where the Amazon River flows into the Atlantic Ocean. Because it is located in a tropical rain forest climate, the days are very warm and humid throughout the year. Afternoon rainfall is very common. The city is located in the largest country in South America.

   The city is   a) Santiago b) Belize c) Recife d) Belém. The country is   a) Paraguay b) Argentina   c) Suriname   d) Brazil.

   Write a question you would like to ask to find out more about this country.

   i. _____

   Human question? _____   Physical question? _____

10. The next city, "J," is located inland, while other large cities in the country are located along the coast. The city is located in the largest country in South America. Because most of the people lived along the coast, the government wanted to encourage people to move away from the coast and settle vast lands inland. To encourage this movement, the capital city was planned and built inland.

   The city is   a) São Paulo b) Brasília c) Rio de Janeiro d) Montevideo. The country is a) Venezuela   b) Argentina   c) Brazil   d) Ecuador.

   Write a question you would like to ask to find out more about this city.

   j. _____

   Human question? _____   Physical question? _____

Name: _____   Date: _____

Unit 11: Where Are You? What Are the Physical Characteristics? What Are the Human Characteristics? (cont.)

MAP 20

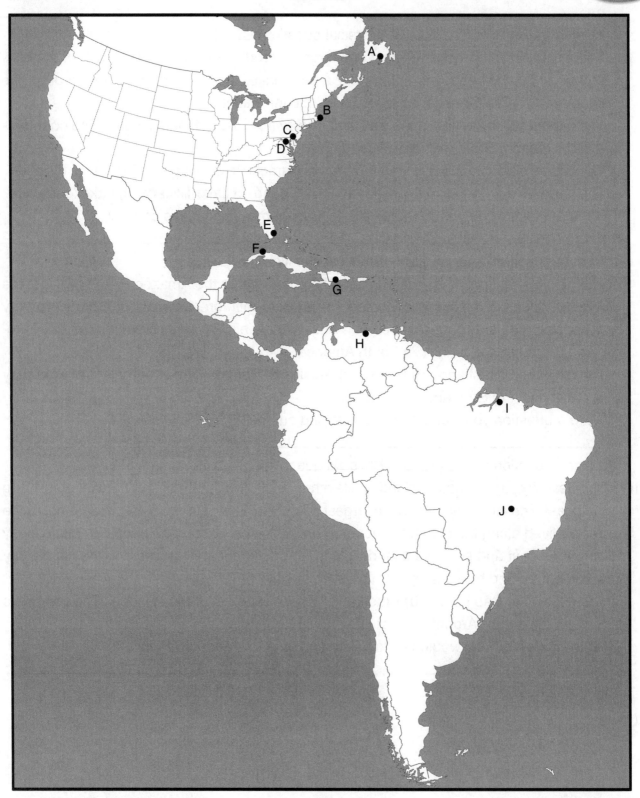

Name: _____   Date: _____

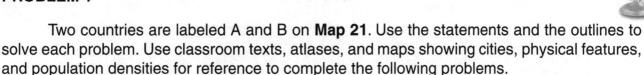

Unit 12: Solve These Problems—#1

PROBLEM 1

Two countries are labeled A and B on **Map 21**. Use the statements and the outlines to solve each problem. Use classroom texts, atlases, and maps showing cities, physical features, and population densities for reference to complete the following problems.

The two countries outlined below are the largest in area on the continents on which each is located. The countries are identified as Country A and Country B.

1. The name of Country A is   a) Canada   b) Argentina   c) United States   d) Brazil.
2. The name of Country B is   a) Canada   b) Argentina   c) United States   d) Brazil.

Use the data below to compare Country A and Country B. If the statement applies to Country A, place a "+" on the blank under A. If the statement applies to Country B, place a "+" on the blank under B. If the statement applies to both countries, place a "+" on both blanks.

		A	B
3.	Contains the largest number of people on the continent on which the country is located.	___	___
4.	Contains the largest area on the continent on which the country is located.	___	___
5.	Most large cities are located near the coast.	___	___
6.	Large cities are located along the coast, and many are located inland.	___	___
7.	Largest river drains a large, sparsely populated region.	___	___
8.	Largest river drains a large, heavily populated region.	___	___
9.	Two largest rivers are important for river transportation.	___	___
10.	Two largest rivers have many major cities located on them.	___	___
11.	Large cities are located along the major rivers.	___	___
12.	Parts of the country have very cold winters.	___	___
13.	Large parts of the country are warm and wet all year.	___	___
14.	Latitude locations north and south of the equator.	___	___
15.	Longitude locations are west of the Prime Meridian.	___	___
16.	The language spoken by most people is Portuguese.	___	___
17.	The language spoken by most people is English.	___	___
18.	Country does not have any neighboring countries to the north.	___	___
19.	Country has a coastal area on the Atlantic and Pacific Oceans.	___	___
20.	Country is bordered by several countries on the north and west.	___	___

21. Summarize how Country A and Country B are alike and different.

_____

_____

_____

_____

National Geography Standard #1: How to use maps and other geographic representations, tools, and technologies to acquire, process, and report information.

Name: _____ Date: _____

Unit 12: Solve These Problems—#1

MAP 21

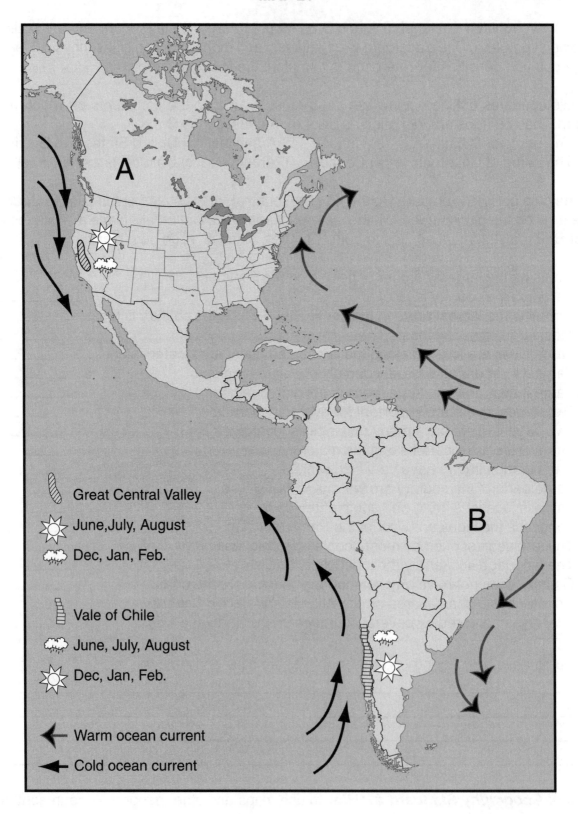

Great Central Valley

June, July, August

Dec, Jan, Feb.

Vale of Chile

June, July, August

Dec, Jan, Feb.

Warm ocean current

Cold ocean current

Name: _____ Date: _____

Unit 12: Solve These Problems—#2

PROBLEM 2

Using **Map 21** and **an atlas**, complete the following. Use the data below to compare the Vale of Chile and the Great Central Valley of California. If the statement applies to the Vale of Chile, place a "+" on the VC blank. If the statement applies to the Great Central Valley, place a "+" on the GC blank. If the statement applies to both, place a "+" on both blanks.

		VC	GC
1.	Summers are warm and dry.	___	___
2.	Winters are wet and mild.	___	___
3.	A warm ocean current is found offshore.	___	___
4.	Located between 30 degrees and 40 degrees north latitude.	___	___
5.	Located between 30 degrees and 40 degrees south latitude.	___	___
6.	Sierra Nevadas are located to the east.	___	___
7.	Andes Mountains are located to the east.	___	___
8.	Located in the Northern Hemisphere.	___	___
9.	Located in the Southern Hemisphere.	___	___
10.	Located in the Western Hemisphere.	___	___
11.	Citrus fruits like oranges and lemons are raised.	___	___
12.	Apples, pears, and peaches are raised.	___	___
13.	A cold ocean current is found offshore.	___	___

14. In your own words, explain why the Great Central Valley of California and the Vale of Chile are similar.

_____

_____

_____

_____

_____

_____

National Geography Standard #4: The physical and human characteristics of places.

Name: _____ Date: _____

Unit 12: Solve These Problems—#3

PROBLEM 3

Use classroom texts, atlases, and maps showing cities, physical features, and population densities for reference to complete the following.

In Brazil, the Amazon River is a large river. In the United States, the Mississippi River is a large river. Both rivers flow many miles and drain a large part of each nation. Place a "+" on the blank for AR if the statement is true for the Amazon River. Place a "+" on the blank MR if the statement is true for the Mississippi River. If the statement applies to both, place a "+" on both blanks.

		AR	MR
1.	The river is the largest river in the country.	____	____
2.	The river is an important transportation route.	____	____
3.	The river flows through fertile farming regions.	____	____
4.	Many cities are located along the river.	____	____
5.	The river flows from north to south.	____	____
6.	The river flows from west to east.	____	____
7.	The river flows through regions with many people.	____	____
8.	The river flows through regions with few people.	____	____
9.	The river flows through a climate that is hot and wet throughout the year.	____	____
10.	The river flows through a region with warm summers and cool winters.	____	____
11.	The river flows through a region that is heavily forested.	____	____
12.	The river flows through a region that is hot and humid throughout the year.	____	____

13. Summarize the major differences between the two rivers.

_____

_____

_____

14. The Amazon River and Mississippi River flow through large geographic regions in each country. Compare the two regions as to how they are alike and different (climate, vegetation, landforms, cities, natural resources, agriculture, communities of people). Use your own paper if you need more room.

_____

_____

_____

National Geography Standard #15: How physical systems affect human systems.

Name: _____ Date: _____

Unit 12: Solve These Problems—#4

PROBLEM 4

The charts below show the monthly temperature and inches of rainfall each month for City A and City B. Use the charts to complete the questions that follow.

City A

Month	J	F	M	A	M	J	J	A	S	O	N	D
Temperature (°F)	78	78	76	77	76	74	74	76	76	77	78	78
Rainfall (inches)	10.0	10.0	12.0	7.0	10.0	7.0	7.0	5.0	9.0	7.0	8.0	11.0

City B

Month	J	F	M	A	M	J	J	A	S	O	N	D
Temperature (°F)	77	77	77	78	80	81	82	82	82	81	79	78
Rainfall (inches)	1.0	0.5	1.0	1.0	4.0	4.0	2.0	4.0	4.0	8.0	3.0	1.0

Compare the yearly temperature range for each city. To compute the yearly temperature range, subtract the lowest monthly temperature from the highest monthly temperature.

1. The yearly temperature range for City A is   a) 10°   b) 20°   c) 8°   d) 4°.

2. The yearly temperature range for City B is   a) 5°   b) 8°   c) 12°   d) 10°.

3. For individuals living in City A and City B, there would be   a) great differences between the cities in yearly temperatures   b) very little difference between the cities in yearly temperatures.

4. The total rainfall for City A is   a) 10   b) 36   c) 103   d) 58   inches.

5. The total rainfall for City B is   a) 20.5   b) 33.5   c) 44.5   d) 12.8 inches.

6. Three consecutive months with the greatest total rainfall for City A are

    a) January, February, March   b) February, March, April

    c) October, November, December   d) June, July, August.

7. Three consecutive months with the greatest total rainfall for City B are

    a) June, July, August   b) December, January, February

    c) August, September, October   d) March, April, May.

8. For individuals living in City A and City B, there would be   a) great difference   b) very little difference   between the cities in yearly rainfall.

9. The greatest difference between City A and City B is in   a) rainfall amount   b) yearly temperatures.

Name: _____ Date: _____

Unit 12: Solve These Problems—#4 (cont.)

10. City A is located in the   a) Northern Hemisphere   b) Southern Hemisphere   because:

_____

_____

_____

11. City B is located in the a) Northern Hemisphere   b) Southern Hemisphere   because:

_____

_____

_____

12. The climate type found in City A is most likely   a) Mediterranean   b) subarctic
    c) tropical rain forest   d) tropical savanna.

Teacher: The student must correctly answer question 12 before continuing.

13. The characteristics of the climate for City A are:

_____

_____

_____

14. The climate type found in City B is most likely   a) Mediterranean   b) subarctic
    c) tropical rain forest   d) tropical savanna.

Teacher: The student must correctly answer Question 14 before continuing.

15. The characteristics of the climate for City B are:

_____

_____

_____

_____

_____

National Geography Standard #1: How to use maps and other geographic representations, tools, and technologies to acquire, process, and report information.

Name: _____ Date: _____

Unit 12: Solve These Problems—#5

PROBLEM 5

Cities of North America and South America

Use classroom texts, atlases, and maps showing cities, physical features, and population densities for reference about North America and South America and complete the following. If the statement applies to North America, place a "+" on the NA blank. If the statement applies to South America, place a "+" on the SA blank. If the statement applies to both, place a "+" on both blanks.

	NA	SA
1. Large cities are located along the east coast.	___	___
2. Few large cities are located inland.	___	___
3. Many large cities are located inland.	___	___
4. Many capital cities are located in the mountains.	___	___
5. Many large cities are located along major rivers.	___	___
6. Few large cities are located along major rivers.	___	___
7. Most large cities on the west coast are located in the mountains.	___	___
8. Few large cities on the west coast are located in the mountains.	___	___
9. Large numbers of people live in large cities along the coasts.	___	___

Use classroom texts, atlases, and maps showing major cities, physical features, and population densities for reference on the United States and Brazil. If the statement applies to the United States, place a "+" on the US blank. If the statement applies to Brazil, place a "+" on the B blank. If the statement applies to both, place a "+" on both blanks.

	US	B
10. Large cities are located along the east coast.	___	___
11. Few large cities are located inland.	___	___
12. Many large cities are located inland.	___	___
13. Many large cities are located along major rivers.	___	___
14. Few large cities are located along major rivers.	___	___
15. Large numbers of people live in large cities along the east coast.	___	___
16. The capital is located inland away from the coast.	___	___
17. The capital is located near other very large cities.	___	___

18. The city of Rio de Janeiro was the capital of Brazil for many years. The government decided to build a new modern capital city and call it Brasília. It was also determined that the new capital should be located inland, away from the coast.

    Most of the large cities and people of Brazil are located near the coast. List some reasons why you think the Brazilian government made the decision to build this new modern capital inland away from the coast and the major population centers. Use your own paper if you need more room.

_____

_____

_____

National Geography Standard #12: The process, patterns, and functions of human settlement.

107

Name: _____   Date: _____

Unit 12: Solve These Problems—#6

PROBLEM 6: Determining how countries are alike and different.

Write the name of the cities below on the blank next to the correct country in which it is located.

Lima   Quito   Bogotá   Caracas   La Paz

Country	**City**
1. Bolivia	_____
2. Peru	_____
3. Ecuador	_____
4. Colombia	_____
5. Venezuela	_____

Place a "+" on the blank if the statement applies to all of the above cities or countries. Place a "–" on the blank if it does not apply to all of the above cities or countries.

_____   6. The language spoken most often is English.

_____   7. The language spoken most often is Spanish.

_____   8. These cities are located in the Andes Mountains.

_____   9. These cities are located on lowlands next to the coast.

_____ 10. These cities are located in the Andes Mountains far from the coast.

_____ 11. These cities are all capitals.

_____ 12. The climate of the cities is the tropical rain forest.

_____ 13. The climate of the cities is the highland climate.

_____ 14. The climate of the cities is the desert climate.

_____ 15. The countries all have port cities located on the ocean.

_____ 16. None of the countries are landlocked.

_____ 17. The countries are very alike in size (area in square miles).

_____ 18. These countries are very alike in number of people (population).

_____ 19. These countries have large areas of plains.

_____ 20. All of the countries border Brazil.

National Geography Standard #12: The process, patterns, and functions of human settlement.

Name: _____ Date: _____

Unit 12: Solve These Problems—#6 (cont.)

Each of the cities below are alike in many ways.

        Lima   **Quito**   **Bogotá**   **Caracas**   **La Paz**

For example, assume that the monthly temperature averages below are characteristic for the above cities.

Answer the following questions.

21. Summarize what the temperature chart below tells you about the climate (temperature range, yearly monthly average, warmest months, coldest months, etc.)

Month	J	F	M	A	M	J	J	A	S	O	N	D
Temperature (°F)	55	56	60	64	65	63	61	61	61	60	57	55

   a. Temperature range yearly: _____

   b. Average annual temperature (to the nearest degree): _____

   c. Warmest Months: _____

   d. Coldest Months: _____

   e. Temperature Difference between the warmest and coldest months: _____

   f. Hemisphere location: a) Northern Hemisphere   b) Southern Hemisphere

22. Explain what kind of clothing you would expect people to wear in this climate.

   _____

   _____

   _____

23. Explain why the five cities might have similar temperatures.

   _____

   _____

   _____

24. Why do you think these cities and the Inca civilization chose the Andes Mountains highlands rather than lower elevations?

   _____

   _____

   _____

National Geography Standard #15: How physical systems affect human systems.

Name: _____ Date: _____

Unit 12: Solve These Problems—#7/#8

PROBLEM 7

1. Place a "+" on the blank if the country has a border on an ocean.

   _____ Argentina   _____ Bolivia   _____ Brazil   _____ Chile
   _____ Colombia   _____ Ecuador   _____ French Guiana   _____ Guyana
   _____ Paraguay   _____ Peru   _____ Suriname   _____ Uruguay
   _____ Venezuela

2. Explain what it means to say that Bolivia and Paraguay are "landlocked" countries.

   _____

   _____

   _____

3. List five reasons why it may be a problem for a country to be landlocked.

   a. _____

   b. _____

   c. _____

   d. _____

   e. _____

National Geography Standard #4: The physical and human characteristics of places.

PROBLEM 8

1. A large oil tanker has just been loaded with oil at a South American port. The tanker will sail from South America into the Caribbean Sea, to a large Northern Hemisphere country that is dependent on other nations for large quantities of oil needed for automobiles, trucks, and manufacturing. It is likely that the oil tanker will be sailing from   a) Uruguay b) Chile c) Venezuela d) Paraguay   to deliver the oil to   a) Canada b) Mexico c) the United States.

2. Assume that for the country in #1 the government changes, and the South American oil supplies are no longer available. The per-gallon price of gasoline in the large Northern Hemisphere nation will now increase, because the supply of oil from South America is no longer available. List five options the large Northern Hemisphere nation should consider to solve the problem of increased prices for oil.

   a. _____

   b. _____

   c. _____

   d. _____

   e. _____

National Geography Standard #11: The patterns and networks of economic interdependence on Earth's surface.

Name: _____ Date: _____

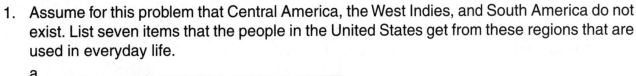

Unit 12: Solve These Problems—#9

PROBLEM 9

1. Assume for this problem that Central America, the West Indies, and South America do not exist. List seven items that the people in the United States get from these regions that are used in everyday life.

   a. _____

   b. _____

   c. _____

   d. _____

   e. _____

   f. _____

   g. _____

2. How would life in the United States be different if the following were no longer available from Central America, the West Indies, and South America?

   a. Fruits: _____

   b. Cacao: _____

   c. Sugar: _____

   d. Bauxite: _____

   e. Coffee: _____

   f. Oil: _____

3. Discuss some ways you think the problem of not having the above items from Central America, the West Indies, or South America could be solved. How would the solution to the problems impact other parts of the world? How would it impact life in the United States?

   _____

   _____

   _____

   _____

   _____

National Geography Standard #11: The patterns and networks of economic interdependence on Earth's surface.

Name: _____ Date: _____

Unit 12: Solve These Problems—#10

PROBLEM 10

1. You are in the United States. Many of the people on the streets speak Spanish rather than English. Many restaurants feature foods with corn, beans, tortillas, and chili. Place a "+" by the four states where this is most likely to occur.

_____ Arizona	_____ California	_____ Idaho
_____ Minnesota	_____ Nebraska	_____ New Mexico
_____ North Dakota	_____ Texas	_____ West Virginia

2. Increasingly, immigrants are coming to the United States from Central America. In the 1970s, the immigrant population from Latin America was approximately 800,000. The immigrant population from Latin America for 2002 is approximately 10,000,000. Many of the new immigrants are from Mexico and Central America. List four reasons that might explain why the immigrant population has increased so dramatically.

a. _____

b. _____

c. _____

d. _____

3. In the 1840s, thousands of Irish immigrants came to the United States. Today, many Hispanics from Mexico and Central America are coming to the United States. Compare the immigration of the Irish and the Hispanics. Why did each group migrate? Where in the United States did they settle? What kinds of jobs did they perform? What cultural characteristics did/do they bring to the United States?

_____

_____

_____

_____

_____

_____

_____

_____

National Geography Standard #9: The characteristics, distributions, and migrations of human populations on Earth's surface.

Name: _____ Date: _____

Unit 12: Solve These Problems—#11

PROBLEM 11

"Goodland" is a country you are getting ready to study.

1. When studying about countries like Goodland, it is important to learn about the human and physical characteristics of the country. Human characteristics might include language, religion, type of government, population, or how people make a living. Physical characteristics might include climate, types of animals, or landforms such as mountains, rivers, and plains. List ten questions that one might ask to learn more about the human and physical characteristics of Goodland.

   a. _____

   b. _____

   c. _____

   d. _____

   e. _____

   f. _____

   g. _____

   h. _____

   i. _____

   j. _____

2. Using a country selected by the teacher, complete the human and physical characteristics for the selected country.

   Human Characteristics:

   a. Population: _____

   b. Religion: _____

   c. Government: _____

   d. Language: _____

   Physical Characteristics:

   a. Rivers: _____

   b. Plains: _____

   c. Climate: _____

   d. Vegetation: _____

National Geography Standard #4: The geographically informed person understands the physical and human characteristics of place.

Name: _____ Date: _____

Unit 12: Solve These Problems—#12

PROBLEM 12

Place a "+" on the blank if the statement would be important in determining the climate that might be found in a given country.

_____ 1. High mountains

_____ 2. Area of the country

_____ 3. Winds that blow toward or away from the country

_____ 4. Latitude location (tropics, mid-latitudes, polar latitudes)

_____ 5. Population of the country

_____ 6. Language spoken in the country

_____ 7. Ocean currents near the country

_____ 8. How the people govern the nation

Map A is a map of the United States showing selected wind belts, ocean currents, and mountains as they presently exist.

Map B is a map of the United States showing the same wind belts and ocean currents as Map A. However, the mountains on Map B have been changed to a new location.

Given—Map A:

1. The winds blow mainly from the west to east.
2. The Rocky Mountains block the west coast from the cold airmasses that blow down from Canada in the winter.
3. Winds blowing down the eastern slope of the Rocky Mountains bring dry conditions to the Great Plains.
4. Warm, moist air from the Gulf of Mexico blows north and meets cooler air from Canada to form storms that bring rain and/or snow to the regions east of the Rocky Mountains.
5. In the eastern part of the United States, winters are cool to cold, with rain and/or snow.
6. In the eastern part of the United States, summers are warm to hot, with rain often falling from thunderstorms.

Problem-solving—Map B

1. What questions would you ask? Presented with the above six statements about Map A, list some questions you would want answered before determining how the new location of the Rocky Mountains might affect the climate on Map B.

    a. _____

    b. _____

    c. _____

    d. _____

Name: _____ Date: _____

Unit 12: Solve These Problems—#12 (cont.)

e. _____

f. _____

Given—Map B:

On Map B, winds still blow generally from west to east. The ocean current off the coast of California is still a cold current flowing south. The cold airmasses from Canada still flow south in winter.

1. What would be a major change with the new location of the Rocky Mountains?

   _____

   _____

2. Describe what you think the climate would be like at "A." Why?

   _____

   _____

3. Describe what you think the winters would be like at "B." Why?

   _____

   _____

4. Describe what you think the summers might be like at "B." Why?

   _____

   _____

5. How would the climate affect agriculture in the region where "A" is located?

   _____

   _____

6. How would the climate affect agriculture in the region where "B" is located?

   _____

   _____

National Geography Standard #15: How physical systems affect human systems.

Unit 12: Solve These Problems—#12 (cont.)

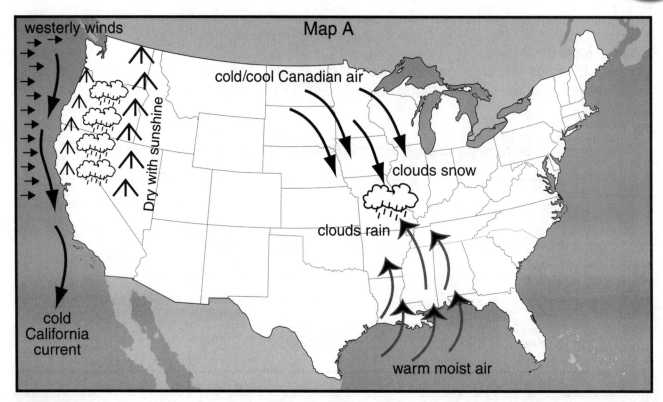

Map A

westerly winds

Dry with sunshine

cold/cool Canadian air

clouds snow

clouds rain

cold California current

warm moist air

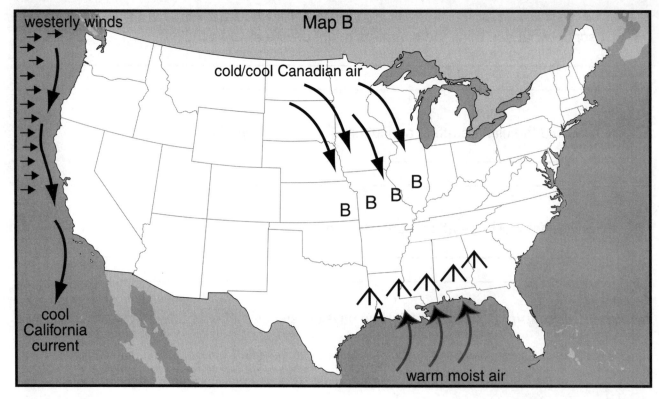

Map B

westerly winds

cold/cool Canadian air

B B B B

A

cool California current

warm moist air

Name: _____   Date: _____

Unit 12: Solve These Problems—#13

PROBLEM 13

Religions in the United States and Brazil

Pie Graph A   **Pie Graph B**

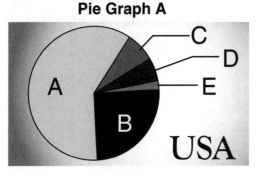   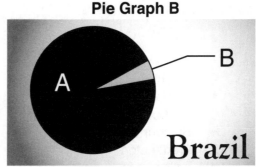

Pie Graph A
A = Protestant
B = Roman Catholic
C = None
D = Other
E = Jewish

Pie Graph B
A = Roman Catholic
B = Other

Using Pie Graph A, complete the following questions.

1. Most people in the United States belong to the   a) Jewish   b) Roman Catholic
   c) Protestant religion.
2. The two largest religious groups are   a) Jewish and other
   b) Roman Catholic and Protestant   c) Protestant and other.
3. The smallest religious group is   a) Jewish   b) Roman Catholic
   c) Protestant.

Using Pie Graph B, complete the following questions.

4. The largest religious group in Brazil is   a) Protestant
   b) Roman Catholic   c) other.
5. If you were traveling in Brazil and you were eating lunch in a restau-
   rant where many Brazilians were having lunch, you would assume
   that   a) most   b) few   c) all
   would be members of the Roman Catholic religion.

Using Pie Graphs A and B, complete the following questions.

6. The religious population of the United States is   a) much like
   b) very unlike   the religious population of Brazil.
7. In your own words, what generalization can you make about the
   religious population in Brazil as compared to the religious population in the United States?

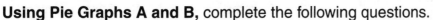

_____

_____

National Geography Standard #10: The character, distribution, and complexity of Earth's cultural mosaics.
National Geography Standard #1: How to use maps and other geographic representations,
tools, and technologies to acquire and process information.

Answer Keys

Unit 1: Political Geography of Western Hemisphere Nations

A. Western Hemisphere (p. 3)
Teacher check Map 1.

B. Canadian Provinces/Territories and Capitals (p. 4–5)
Teacher check Map 2.
Pretest Practice

1. Victoria		2. Toronto	
3. Charlottetown		4. Edmonton	
5. Quebec		6. Halifax	
7. Regina		8. St. John's	
9. Winnipeg		10. Fredericton	
11. Whitehorse		12. Yellowknife	
13. Iqaluit			

B. Canadian Provinces/Territories and Capitals—Test (p. 6)

1. b	2. a	3. d	4. a
5. a	6. b	7. d	8. d
9. d	10. c	11. a	12. d
13. c			

C. United States and Capitals (p. 7–8)
Teacher check Map 3.

C. United States and Capitals—Pretest Practice (p. 9–10)

1. Juneau	2. Honolulu
3. Austin	4. Sacramento
5. Tallahassee	6. Salem
7. Olympia	8. Salt Lake City
9. Boise	10. Phoenix
11. Santa Fe	12. Carson City
13. Helena	14. Cheyenne
15. Denver	16. Bismarck
17. Pierre	18. Lincoln
19. Topeka	20. Oklahoma City
21. Little Rock	22. Baton Rouge
23. Jefferson City	24. Des Moines
25. St. Paul	26. Springfield
27. Nashville	28. Frankfort
29. Jackson	30. Madison
31. Indianapolis	32. Columbus
33. Montgomery	34. Atlanta
35. Columbia	36. Raleigh
37. Richmond	38. Annapolis
39. Dover	40. Trenton
41. Harrisburg	42. Albany
43. Providence	44. Hartford
45. Boston	46. Montpelier
47. Concord	48. Augusta
49. Charleston	50. Lansing
51. San Juan	

C. United States and Capitals—Test (p. 11–12)

1. b	2. a	3. d	4. b
5. d	6. c	7. b	8. a
9. c	10. d	11. c	12. b
13. a	14. d	15. b	16. a
17. c	18. d	19. b	20. d
21. a	22. b	23. a	24. c
25. d			

D. Central American Countries and Capitals (p. 13)
Teacher check Map 4.
Pretest Practice

1. Belmopan	2. San José
3. Panamá	4. Tegucigalpa
5. Managua	6. Mexico City
7. San Salvador	8. Guatemala

D. Central American Countries and Capitals—Test (p. 14)

1. c	2. a	3. b	4. a
5. b	6. c	7. c	8. d

E. West Indies Countries and Capitals (p. 15)
Teacher check Map 5.
Pretest Practice

1. Havana	2. Port-au-Prince
3. Kingston	4. Santo Domingo
5. San Juan	6. Port of Spain

E. West Indies Countries and Capitals—Test (p. 16)

1. c	2. a	3. d	4. a
5. b	6. a		

F. South American Countries and Capitals (p. 17–18)
Teacher check Map 6.
Pretest Practice

1. Caracas	2. Santiago
3. Bogotá	4. Quito
5. Brasília	6. Montevideo

7. Asunción   8. Lima
9. La Paz   10. Paramaribo
11. Cayenne   12. Georgetown
13. Buenos Aires

F. South American Countries and Capitals—Test (p. 19)

1. b   2. d   3. a   4. d
5. c   6. a   7. b   8. b
9. c   10. a   11. c   12. d
13. b

G. Identifying and Locating Countries (p. 20)

1. A. Puerto Rico   B. West Indies
2. A. Paraguay   B. South America
3. A. United States   B. North America
4. A. Chile   B. South America
5. A. Canada   B. North America
6. A. Panama   B. Central America
7. A. Argentina   B. South America
8. A. Mexico   B. Central America
9. A. Colombia   B. South America
10. A. Guatemala   B. Central America
11. A. Peru   B. South America
12. A. Brazil   B. South America

Unit 2: Population and Area of Western Hemisphere Nations
Rankings (p. 21–22)

A.	Area	B.	Population
1.	Canada	1.	United States
2.	United States	2.	Brazil
3.	Brazil	3.	Mexico
4.	Argentina	4.	Colombia
5.	Mexico	5.	Argentina
6.	Peru	6.	Canada
7.	Colombia	7.	Peru
8.	Venezuela	8.	Venezuela
9.	Chile	9.	Chile
10.	Paraguay	10.	Ecuador
11.	Ecuador	11.	Cuba
12.	Guyana	12.	Dominican Republic
13.	Uruguay	13.	Haiti
14.	Cuba or Honduras	14.	El Salvador or Honduras
15.	Cuba or Honduras	15.	El Salvador or Honduras
16.	Panama	16.	Paraguay
17.	Costa Rica	17.	Costa Rica
18.	Dominican Republic	18.	Uruguay
19.	Haiti	19.	Panama
20.	El Salvador	20.	Jamaica
21.	Jamaica	21.	Trinidad and Tobago
22.	Trinidad and Tobago	22.	Guyana

C. Comparing the Areas of Countries (p. 23)

1. Canada   2. United States
3. Brazil   4. Argentina
5. Mexico   6. Peru
7. Colombia   8. Venezuela
9. Chile   10. Paraguay

North America: Canada, United States
Central America: Mexico
South America: Brazil, Argentina, Peru, Colombia, Venezuela, Chile, Paraguay

D. Comparing the Populations of Countries (p. 23)

1. United States   2. Brazil
3. Mexico   4. Colombia
5. Argentina   6. Canada
7. Peru   8. Venezuela
9. Chile   10. Ecuador

North America: United States, Canada
Central America: Mexico
South America: Brazil, Colombia, Argentina, Peru, Venezuela, Chile, Ecuador

D. Comparing the Populations of Countries—Pie Graph A (p. 24)

1. b   2. c   3. b   4. c
5. d   6. a   7. a   8. b
9. c

Pie Graph B (p. 25)

1. b   2. c   3. c   4. b
5. c   6. a

E. Comparing the Populations and Areas of Countries—Pretest Practice (p. 26)

1. Brazil   2. Canada
3. Brazil   4. the United States
5. Canada   6. the United States
7. Brazil, Argentina, Peru
8. Brazil, Colombia, Argentina
9. Canada, the United States, Brazil
10. the United States, Brazil, Mexico

11.   Honduras
12.   Honduras, El Salvador
13.   Cuba
14.   Cuba

E. Comparing the Populations and Areas of Countries—Test (p. 27)
1. b   2. c   3. d   4. a
5. a   6. b   7. b   8. b
9. a   10. b   11. a   12. b
13. c   14. c

Unit 3: Physical Features of the Western Hemisphere
A. Rivers (p. 28)
a. Mississippi   b. Missouri
c. Ohio   d. Tennessee
e. Columbia   f. Snake
g. Potomac   h. Hudson
i. Platte   j. Fraser
k. Rio Grande   l. San Joaquin
m. Sacramento   n. Yukon
o. St. Lawrence   p. Colorado
q. Illinois   r. Mackenzie
s. Paraná   t. Orinoco
u. Magdalena   v. Amazon

A. Rivers—Crossword Puzzle (p. 29)

B. Mountains and Volcanoes
Map Activity (p. 30–31)
1.   a. Sierra Nevadas   b. Catskills
     c. Adirondack   d. Cascades
     e. Rockies   f. Appalachian
     g. Andes   h. Brazilian
                                  Highlands
     i. Alaskan Range   j. Brooks Range
     k. Blue Ridge

2.   a. Mt. McKinley   b. Mt. Aconcagua
     c. Mt. Cotopaxi   d. Mt. Chimborazo
     e. Mt. Popocatépetl   f. Mt. Izttaccíhuatl
     g. Mt. Logan   h. Mt. Whitney
     i. Mt. Shasta   j. Mt. Rainier
     k. Pikes Peak

Rankings and Graph (p. 32)
3.   a. Mt. Aconcagua; 22,800; Argentina; SA
     b. Mt. Chimborazo; 20,700; Ecuador; SA
     c. Mt. McKinley; 20,300; United States; NA
     d. Mt. Logan; 19,600; Canada; NA
     e. Mt. Cotopaxi; 19,300; Ecuador; SA
     f. Mt. Popocatépetl; 17,900; Mexico; NA
     g. Mt. Iztaccíhuatl; 17,200; Mexico; NA
     h. Mt. Whitney; 14,500; United States; NA
     i. Mt. Rainier; 14,400; United States; NA
     j. Mt. Shasta; 14,200; United States; NA
     k. Pikes Peak; 14,100; United States; NA

C. Plains and Valleys (p. 33–34)
1. b   2. b   3. c   4. a
5. a   6. b   7. b

D. Plateaus (p. 33–34)
1. a   2. c   3. a   4. b
5. c

E. Lakes, Seas, Gulfs, and Bays (p. 35–36)
1. b   2. b   3. a   4. c
5. b   6. a   7. c   8. a
9. a   10. b   11. c   12. b
13. b
Teacher check Map 10.

F. Physical Features—Pretest Practice (p. 37)
1.   Mississippi   2.   Amazon
3.   Rio Grande   4.   Mackenzie
5.   Aconcagua   6.   Argentina
7.   McKinley   8.   California
9.   Pampas   10.   Altiplano
11. Great Plains   12.   Gran Chaco
13. Bolivia, Peru   14.   Hudson
15. Caribbean   16.   Chesapeake

F. Physical Features—Test (p. 38)
1. c   2. a   3. b   4. b
5. a   6. d   7. c   8. c
9. d   10. b   11. a   12. a
13. c   14. b   15. c   16. d

Unit 4: Using Latitude and Longitude
Diagram 1 (p. 39–40)

1. Teacher check.
2. Teacher check.
3. c
4. b
5. a, b
6. Teacher check.
7. c
8. a
9. b, a
10. 15°N, 150°W
11. 45°N, 90°W
12. 60°N, 0°
13. 15°S, 60°W
14. 30°N, 60°E
15. 45°S, 90°E
16. 0°, 150°E

Map Activity—Map 11 (p. 40–41)

1. b
2. a
3. c
4. b
5. a
6. b
7. a
8. b
9. N: Mexico, Canada, Panama, Cuba, United States
   S: Bolivia, Chile, Peru, Uruguay

Unit 4: Using Latitude and Longitude—Pretest Practice (p. 42)

1. equator
2. Prime Meridian
3. equator
4. Prime Meridian
5. 180
6. parallel
7. Poles
8. 90
9. Northern
10. Southern
11. Eastern
12. Western
13. Northern, Western, north
14. Southern, Eastern, east

Unit 4: Using Latitude and Longitude: Test (p. 43)

1. b
2. a
3. b
4. a
5. d
6. c
7. c
8. c
9. a
10. c
11. d
12. b
13. b, c
14. a, d

Unit 5: Climate in the Western Hemisphere
A. Tundra Climate (p. 44)
Teacher check Map 12a.

1. 40
2. -20
3. b
4. c
5. Newfoundland, Quebec, Yukon Territory, Nunavut Territory, Northwest Territories
6. Iqaluit

B. Subarctic Climate (Taiga) (p. 46)
Teacher check Map 12a.

1. c
2. Alberta, Newfoundland, Northwest Territories, Ontario, Quebec, Nunavut Territory
3. b
4. c
5. c
6. b

C. Humid Continental Climate (p. 47)
Teacher check Maps 12a, 12b, and 12c.

1. b
2. Alberta, Nova Scotia, Ontario, Quebec, Saskatchewan
3. Maine, Minnesota, New York, Ohio, Vermont, Wisconsin
4. a
5. b
6. True

D. Humid Subtropical Climate (p. 48–49)
Teacher check Map 12a.

1. c
2. a
3. b
4. a
5. c, a
6. Alabama, Georgia, North Carolina, South Carolina, Virginia
7. Argentina, Uruguay
8. c
9. a
10. a
11. a

E. Steppe or Semiarid Climate (p. 51)
Teacher check Maps 12a, 12b, and 12c.

1. Teacher check.
2. c
3. True
4. True
5. False
6. False
7. True
8. True

F. Desert Climate (p. 52)

1. Arizona, New Mexico, Nevada
2. Chile, Argentina
3. c
4. a
5. c

G. Highland Climate (p. 52–53)
Diagram 4

1. b
2. d

Map Activity

1. Teacher check Maps 12b and 12c.
2. b
3. c
4. Colorado, Idaho, Oregon, Washington, Wyoming

H. Mediterranean Climate (p. 54–55)

1. Teacher check.
2. Chile, South America
3. Teacher check.
4. California, North America
5. a
6. a
7. c
8. c
9. a, b
10. b, a

Map Activity

1. Teacher check.
2. west
3. 30, 40
4. 30, 40
5. b
6. c
7. b
8. c
9. b
10. a
11. d

I. West Coast Marine Climate (p. 56–57)
Diagram 5
1. b

Map Activity
1. Teacher check maps.   2. c
3. b   4. a   5. b
Teacher check graph.

J./K. Tropical Rain Forest Climate/Tropical Savanna Climate (p. 58–59)
Map Activity
1–4. Teacher check Map 12c.
5. c   6. a   7. a   8. b
9. b   10. a   11. c   12. a

L. Identifying Climates (p. 60)
1. North America: desert, highland, humid continental, humid subtropical, Mediterranean, steppe, subarctic, tropical savanna, tundra, west coast marine
   South America: desert, highland, humid subtropical, Mediterranean, steppe, tropical rain forest, tropical savanna, west coast marine
2. b   3. c   4. c

M. Climate—Pretest Practice (p. 61)
1. highland   2. tundra
3. humid subtropical   4. Mediterranean
5. tropical rain forest 6. west coast marine
7. Peru   8. California
9. tropical rain forest 10. orographic
11. three   12. eastern

M. Climate—Test (p. 62)
1. a   2. c   3. d   4. a
5. b   6. b   7. a   8. d
9. d   10. a   11. a   12. c

Unit 6: Major Cities of the Western Hemisphere Map Activity/Chart (p. 63–64)
Teacher check Map 13.
1. Mexico City, North America, inland
2. Seattle, North America, coastal
3. Boston, North America, coastal
4. Caracas, South America, inland
5. Bogotá, South America, inland
6. Buenos Aires, South America, coastal
7. Detroit, North America, inland
8. Buffalo, North America, inland
9. Philadelphia, North America, inland
10. Washington, D.C., North America, coastal
11. Cleveland, North America, inland
12. Baltimore, North America, coastal
13. São Paulo, South America, coastal
14. Guayaquil, South America, coastal
15. Belo Horizonte, South America, inland
16. Miami, North America, coastal
17. New Orleans, North America, coastal
18. Houston, North America, coastal
19. Chicago, North America, inland
20. St. Louis, North America, inland
21. San Francisco, North America, coastal
22. Milwaukee, North America, inland
23. Minneapolis, North America, inland
24. Denver, North America, inland
25. Dallas, North America, inland
26. Los Angeles, North America, coastal
27. Kansas City, North America, inland
28. Cincinnati, North America, inland
29. Belém, South America, coastal
30. Recife, South America, coastal
31. Medellín, South America, inland
32. Montreal, North America, inland
33. Toronto, North America, inland
34. New York City, North America, coastal

Unit 6: Major Cities—Inland/Coastal and Graph (p. 65)
1. a. I b. I c. C d. I e. I f. I
   g. C h. C i. C j. I k. C l. I
   m. C n. C o. C p. C
2. a   3. c
4. Teacher check bar graph.
5. b   6. c   7. a

Unit 6: Major Cities—Pretest Practice (p. 66–67)
1. Lima   2. Chicago
3. Washington, D.C.   4. Toronto
5. Minneapolis   6. New York City
7. Buffalo   8. Montreal
9. Cincinnati   10. Houston
11. Kansas City   12. New Orleans
13. Miami   14. Los Angeles
15. San Francisco   16. Milwaukee
17. Philadelphia   18. Mexico City
19 Bogotá   20. Caracas
21. St. Louis   22. Detroit
23. Denver   24. Boston
25. Seattle   26. Belém
27. Rio de Janeiro   28. Buenos Aires
29. Recife   30. Cleveland
31. Baltimore   32. São Paulo
33. Guayaquil   34. Belo Horizonte

Unit 6: Major Cities—Test (p. 68)

1. a   2. a   3. b   4. c
5. d   6. a   7. b   8. d
9. a   10. c   11. a   12. c
13. a   14. b   15. a   16. d
17. c   18. d   19. c   20. a

Unit 7: Agriculture in the Western Hemisphere
A. Tropical Crops—Map Activity (p. 69–70)

1. Teacher check Map 14.
2. Brazil, Costa Rica, Cuba, Dominican Republic, El Salvador, Guatemala, Haiti, Honduras, Jamaica, Panama, Puerto Rico, Trinidad
3. Cuba, Dominican Republic, Jamaica, Puerto Rico, Trinidad
4. bananas, coffee, oranges, rice, sugar cane, tobacco
5. Colombia, Costa Rica, Cuba, Haiti, Honduras, Jamaica, Nicaragua

A. Tropical Crops—Pretest Practice (p. 71)

1. $23\frac{1}{2}°$, $23\frac{1}{2}°$ 2. bananas
3. West Indies   4. Central America
5. coffee   6. Cancer
7. Capricorn   8. rice
9. cacao   10. Trinidad
11. rain forest   12. savanna
13. highland   14. Honduras
15. Caribbean   16. tropics
17. sugar cane

A. Tropical Crops—Test (p. 72)

1. a, c   2. c   3. d   4. b
5. a   6. a   7. b   8. c
9. b   10. c   11. d   12. c
13. a   14. c   15. c   16. b
17. a

B. Mid-latitude Crops—Map Activity 1 (p. 73)

1. Teacher check Map 14.
2. Argentina, Canada, Chile, United States, Uruguay
3. Argentina, Canada, Chile, United States, Uruguay
4. apples, corn, cotton, grapefruit, oats, oranges, pears, soybeans, tobacco, wheat

B. Mid-latitude Crops—Map Activity 2 (p. 73–75)

1–3. Teacher check Map 15a.
4. a   5. b   6. c

7. Teacher check Map 15a. Alabama, Florida, Georgia, North Carolina, Louisiana, Mississippi, South Carolina

Map Activity 3 (p. 74–75)

1. Teacher check Map 15b.
2. b   3. c   4. a
5–7. Teacher check Map 15b.

B. Mid-Latitude Crops—Pretest Practice (p. 76)

1. 30, 60
2. apples, corn, cotton, oranges, pears, wheat
3. beef, butter, chicken, milk, pork
4. Argentina, Canada, Chile, United States, Uruguay
5. Great Central Valley
6. Vale of Chile
7. Ontario
8. Great Plains
9. Pampas
10. Corn Belt
11. Willamette

B. Mid-Latitude Crops—Test (p. 77)

1. a   2. c   3. c   4. b
5. c   6. d   7. a   8. a
9. c   10. c

Unit 8: Natural Resources of the Western Hemisphere (p. 78–80)
A. Oil

1. Canada, Mexico, United States, Venezuela
2. c   3. a   4. b

B. Iron Ore

5. Brazil, Canada, Venezuela
6. b   7. b

C. Copper

8. a   9. a   10. b

D. Bauxite

11. c   12. c

E. Coal

13. b

F. Forests

14. d   15. d   16. a   17. d

Unit 8: Natural Resources—Pretest Practice (p. 81)

1. Venezuela   2. Trinidad
3. Maracaibo Basin   4. Labrador/ Newfoundland

5. Chile
6. Minas Gerais
7. Mesabi Range
8. El Pao
9. Bingham Canyon
10. Canadian Shield
11. oil
12. copper
13. Jamaica
14. coal
15. lumber
16. Canada/United States

Unit 8: Natural Resources—Test (p. 82)
1. c 2. a 3. d 4. a
5. c 6. a 7. c 8. a
9. b 10. b 11. a 12. a
13. b 14. a 15. a 16. c

Unit 9: Central America and the West Indies
A. Central America—Climate (p. 83–84)
1. Belize, Costa Rica, El Salvador, Guatemala, Honduras, Nicaragua, Panama
2. a 3. b 4. b 5. a
6–8. Teacher check.
9. Guatemala, Honduras, Nicaragua
10. b
11. Belize, Costa Rica, Guatemala, Honduras, Nicaragua, Panama
12. a

A. Central America: Agriculture (p. 84)
1. a 2. a 3. a
4. Costa Rica, Honduras, Nicaragua, Panama

B. West Indies: Capital Cities (p. 85)
1–2. Teacher check Map 19.

B. West Indies: Climate (p. 85)
1. b 2. b 3. True 4. True

B. West Indies: Agricuture (p. 85)
1. a 2. c 3. b 4. b
5. a

C. Pretest Practice (p. 87)
Central America:
1. isthmus 2. warm
3. east 4. onshore
5. wet 6. west
7. offshore 8. wet, dry
9. onshore 10. offshore
11. bananas, coffee, sugar cane
12. coffee

West Indies:
1. sugar cane 2. cacao
3. coffee 4. bananas
5. tobacco 6. tropical savanna
7. summer 8. temperature

C. Central America and the West Indies—Test (p. 88)
1. c 2. a 3. c 4. b
5. a 6. a 7. a 8. c
9. a 10. b 11. b 12. d
13. a 14. c 15. a 16. d
17. c 18. b 19. a 20. b

Unit 10: Which Country, State, Province, or Territory Is Described?
A. (p. 89–91)
1. a) Mexico, b) Northern, c) the United States
2. a) Venezuela, b) Northern, c) Brazil
3. a) the Dominican Republic, b) Hispaniola, c) Haiti
4. a) California, b) Pacific, c-d) Los Angeles, San Diego
5. a) Canada, b) Nova Scotia, c) Halifax, d) New Brunswick
6. a) Arizona, b) the United States, c) Phoenix
7. a) Chile, b) Santiago, c) Andes, d) South America, e) Argentina
8. a) Brazil, b) Brasília
9. a) El Salvador, b) San Salvador, c) Central
10. a) Paraguay, b) Asunción, c) Gran Chaco

B. (p. 92–93)
1. a) Bolivia, b-c) La Paz, Sucre, d-e) Poopó, Titicaca
2. a) Prince Edward Island, b) Charlottetown, c) Canada, d) Gulf of St. Lawrence
3. a) Cuba, b) Havana, c) Gulf of Mexico, d) Caribbean
4. a) Ecuador, b) Quito, c) Galápagos, d) Guayaquil
5. a-b) Oregon, Washington c-d) Salem, Olympia, e) Columbia, f) Cascades
6. a-b) Arkansas, Missouri, c-d) Little Rock, Jefferson City, e) the United States, f) St. Louis, g) Arkansas, h) Iowa, i) Louisiana

Unit 10: Which Country, State, Province, or Territory is Described?: Crossword Puzzle—Canada (p. 94)

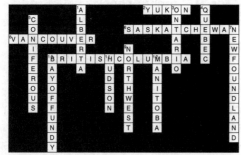

Unit 10: Which Country, State, Province, or Territory is Described?: Crossword Puzzle—United States (p. 95)

Unit 10: Which Country, State, Province, or Territory is Described?: Crossword Puzzle—Central and South America (p. 96)

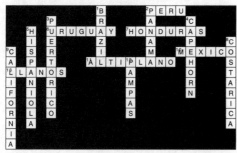

Unit 11: Where Are You? What are the Physical Characteristics? What Are the Human Characteristics? (p. 97–100)

1. b, d, teacher check
2. a, b, teacher check
3. d, b, teacher check
4. a, teacher check
5. d, c, teacher check
6. a, c, teacher check
7. d, c, teacher check
8. c, a, d, teacher check
9. d, d, teacher check
10. b, c, teacher check

Unit 12: Solve These Problems
PROBLEM 1 (p. 101–102)

1. a	2. d	3. B	4. A, B
5. B	6. A	7. B	8. A
9. A, B	10. A	11. A	12. A
13. B	14. B	15. A, B	16. B
17. A	18. A	19. A	20. B

21. Teacher check.

PROBLEM 2 (p. 103)

1. VC, GC	2. VC, GC	3. Neither
4. GC	5. VC	6. GC
7. VC	8. GC	9. VC
10. GC, VC	11. GC, VC	12. GC, VC
13. GC, VC	14. Teacher check.	

PROBLEM 3 (p. 104)

1. AR, MR	2. AR, MR	3. AR, MR
4. MR	5. MR	6. AR
7. MR	8. AR	9. AR
10. MR	11. AR	12. AR, MR

13–14. Teacher check.

PROBLEM 4 (p. 105–106)

1. d	2. a	3. b	4. c
5. b	6. a	7. c	8. a

9. a
10. b, The lowest temperatures are in June and July, indicating winter is during these months.
11. a, The lowest temperatures are in January and February, indicating winter is in these months.
12. c
13. High monthly average temperatures and rainfall throughout the year, vegetation is dense forest
14. d
15. High monthly average temperatures, definite wet and dry seasons, vegetation is tall grass and shrubs or small trees

PROBLEM 5 (p. 107)

1. NA, SA	2. SA	3. NA
4. SA	5. NA	6. SA
7. SA	8. NA	9. NA, SA
10. US, B	11. B	12. US
13. US	14. B	15. US, B
16. B	17. US	

18. Because Brazil is a large country with much sparsely settled land away from the coast, an effort was made to get the population to move inland where large amounts of land are available for farming.

PROBLEM 6 (p. 108–109)
1. La Paz   2. Lima   3. Quito
4. Bogotá   5. Caracas
6. –   7. +   8. +
9. –   10. –   11. +
12. –   13. +   14. –
15. –   16. –   17. –
18. –   19. –   20. –
21a. 10°   b. 60°
  c. April, May, June
  d. December, January, February
  e. 10°   f. a
22. Sweaters, long pants, light jackets, hats
23. All are located in mountains with a highland climate.
24. Plateaus in the mountains had a much milder climate than in lowlands.

PROBLEM 7 (p. 110)
1. Argentina, Brazil, Chile, Colombia, Ecuador, French Guiana, Guyana, Peru, Suriname, Uruguay, Venezuela
2. They do not have access to an ocean or sea. Each country must gain access to an ocean by crossing a neighboring country.
3. Teacher check. Transportation and movement issues should be discussed.

PROBLEM 8 (p. 110)
1. c, c
2. Teacher check. Conservation and alternative energy options should be discussed.

PROBLEM 9 (p. 111)
1–3. Teacher check. Answers will vary.

PROBLEM 10 (p. 112)
1. Arizona, Califronia, New Mexico, Texas
2–3. Teacher check. Answers will vary.

PROBLEM 11 (p. 113)
1–2. Teacher check. Answers will vary.

PROBLEM 12 (p. 114–116)
Plus signs should be on 1, 2, 3, 4, 7.
Problem-solving—Map B: Teacher check.
Given—Map B: Teacher check.

PROBLEM 13 (p. 117)
1. c   2. b   3. a   4. b
5. a   6. b
7. In Brazil, most people have very similar religious beliefs; they worship in the same manner, etc. In the United States, there is a wider variety of religious beliefs and practices.